Make Energy Cool

A Blueprint to Make the Future of Energy Our Own

By Sagi Aloni

"Believe nothing, no matter where you read it, or who said it, no matter if I have said it, unless it agrees with your own reason and your own common sense."

Gautama Buddha

To David Bowie, who inspired us all to be heroes

Author: Sagi Aloni

Editor: Cheri Hanson

Cover Art: Liat Aloni

Table of Contents

I. PREFACE

This book began as a series of LinkedIn posts on New Year's Eve 2015. The network asked its members to share their big ideas for 2016 under the #BigIdeas2016 meta-search. It then collated the posts into a single list of Big Ideas, which gave members a look at cross-industry trends and predictions from the world's top minds and influencers. I wanted to challenge myself to create a new, exciting idea for 2016.

The LinkedIn challenge coincided with a time of great change in our energy landscape. Earlier that year, world oil prices had begun to decline rapidly until, in early 2016, they reached a 12-year low.

On December 12, 2015, the leaders of nearly 200 nations gathered in Paris and signed what was probably the environmental movement's greatest achievement to date, *The New Climate Treaty*, or the *COP 21*.

And so, with 2015 about to end, I tried to figure out my Big Idea for 2016. I realized that my two strongest passions, a concern for the environment and my love for creativity, have rarely met. I have been drawing, writing, designing, cooking, and experimenting with creative mediums for as long as I can remember. I consume an endless stream of books, comics, TV shows, movies and art, and I adore creative people of all fields.

My concern for the environment has led me to work as a project manager and consultant for large-scale infrastructure projects for more than a decade and across five continents. I've participated in stationary energy projects of various scales and resources, including renewables, hydropower and natural gas projects. Typically, I have worked with or led professional planning and design teams who aim to deliver large projects with minimal environmental impacts, a manageable public opposition, and quick regulatory approvals.

This professional role has required me to be more serious and overcome more tedious challenges than I'm naturally inclined to tackle. Because some of my responsibilities included conveying these projects to the general public, environmental activists, and regulators, I have learned the hard way that energy technology and policy have become extremely complicated, delicate issues. In their early stages, energy projects are often secretive and influenced by intense political negotiations. When they are finally revealed to the public, *everybody* has something to say about them and the public response is always negative.

Whenever I had to work on an energy project, I realized that my greatest efforts were spent not on my job, but rather on ensuring I didn't say the wrong thing or make the wrong assumptions, in order not to upset anyone. In short, I found energy neither fun nor creative.

So this was my big idea: Can energy be cool?[1] Can it be relevant, fun and transparent? Can we treat electric energy technology the same way we treat other technologies, like cars, computers, airplanes, and electronics? Can we get excited about energy generation technology, instead of being afraid of it? Is the current energy policy and its sacred cow of Energy Stability the best way forward?

It struck me as odd that in 2015, our electric energy came from technologies that were between 50 and 150 years old, while so many of the technologies we use daily are 2-3 years old, and sometimes even newer than that.

---

[1] Just so we're clear, I have no idea what *makes* something cool. I was never a trendsetter, nor have I ever marketed stuff to teenagers, who always seem to know what is currently cool. I just know that something cannot be dull and uninspiring and cool at the same time.

And so I published the first Make Energy Cool[2] post. At first, there was no reaction. Then came the first comment, and then another, and then the views started to add up to three digits, then four, and they just kept growing. But getting the traffic was just a small part of the story. The LinkedIn navigation bar allows a writer to see where all his or her traffic is coming from – and it came from everywhere: the U.S., the U.K., Australia, Kenya, India, Iran, the Netherlands, France, Pakistan, Russia, Israel, and more. Some of the readers worked in energy and some were students, while others worked in public relations, consulting, and administration. And somehow this idea has generated substantial interest in many different places and among many different people. Reading the comments made me think that people everywhere are concerned about our current energy landscape and its implications for our future, but fresh ideas for change are in short supply. Many comments focused on how we keep framing the energy/climate debate in isolation, without exploring what is happening in other fields and learning how to make energy better.

While I never intended that post to be the next Gangnam Style (which is funnier *and* cooler), I was excited to see a sizeable reaction, because it meant that my idea was not so outrageous. That it was, in fact, possible.

---

[2] It's still there, look it up under my name.

The nano-global response on LinkedIn encouraged me to write this book, but its timing isn't just about the end of one year, nor the result of a few hundreds likes (though I cherish each one). This is not a popularity contest. Millions of people spend more time and mental energy following the Kardashian family's exploits than exploring energy issues.

The main purpose of this book is to challenge our current energy debate, to help us understand it better, and more importantly, to highlight what we can do to create our own energy future. At this point, it is also important to clarify that when I talk about energy, I am in fact talking about *energy generation*, or the production of electric energy. Mobile energy for transportation is a totally different story and faces many other equally important issues, which sadly won't be addressed in this book[3].

To paraphrase Winston Churchill, the COP 21 agreement is *not the end of something. It is not even the end of the beginning.* Our energy landscape will change in the next 2-3 decades, whether we welcome it or not. Current estimates predict that we will need around 40 percent increased energy

---

[3] Though Make Energy Cool Version 2.0 is always an option.

capacity during this time, which is similar to adding a new China and a new U.S. into the global grid by 2050.

This growth can be attributed to world population growth, but also due to a rise in the global standard of living, as electric energy replaces other power sources and automates many other technologies and human efforts.

At the same time, the COP 21 intends to reshape the global energy mix, the technologies we use to generate energy, and to create a more transparent international system of energy and climate information. In an ideal world, this would mean that we would all agree on where our energy should come from, how it's generated, how it's distributed, and how it can have the lowest possible negative impact on human lives, on economic growth, and on our common home planet. But in the real world, energy is a contentious issue that is often secretive, and surrounded by the conflicting interests of nations, corporations and powerful individuals.

The convergence of these two influences – a quick restructuring of the global energy landscape and the complicated language, politics and war of interests that shape our current system – reminded me of the wise words of Albert Einstein, who described insanity as "the outcome of doing the same thing over and over again and expecting different results."

If the best thing we can do, as global citizens, to prevent unprecedented climate change is to expect our governments and energy giants to help us avoid this type of collective insanity, then we are all delusional. As I will demonstrate later, we made a similar assumption in the past and it didn't turn out as we expected.

There's a good chance that at least some of the ideas presented in the book can be better articulated, explained, or justified. But I challenge you to come forward with as many innovative and creative ideas as you can to envision a new energy future. Sadly, our current energy landscape has changed little over the last 30-40 years, and the debate over the future of our energy has been similarly stagnant across three or four decades.

In the words of Khaleesi from the *Game of Thrones* series, *it's time to break the whe*el. But unlike Khaleesi's plans, which are a figment of author George R. R. Martin's imagination, the future of energy will require the collective effort of many, and we all have a say here. If this book inspires even one person to come up with a new technology, a new business model or an inspiring, creative project that will change the way energy is perceived, then it's worth every moment of my time (and yours), and all the scrutiny in the world.

Let's Make Energy Cool, together.

Sagi

January 2016

II. WHAT'S IN THIS BOOK?

This book is all about making *energy generation* relevant to *the current generation*. Consumer energy, or electricity, doesn't have bad public relations nor is there a lack of demand; it is one of the most desired services on the planet, and everybody wants it, day or night – most urgently when their mobile battery is about to die. But while we love to *consume* energy, we seem to ignore and often hate the systems (and sometimes the people) that *generate* energy.

I want to assess how we've come to fear and loathe something we deeply want and constantly use, but I also want to offer more than just a rear view analysis; there are many books that tackle the history and evolution of the energy industry and the formation of energy policies. I'll touch on some key moments in the history of mass energy production, but this is not an accurate, nor a linear historical account.

With your permission, I will also leave out the heated climate change debate that has dominated the last decade. I think we are beyond that point, and there are better and more knowledgeable writers on this subject. Nor have I tried to explain the growing environmental and financial impacts of central energy generation and distribution, for similar reasons.

I also wrote this book from the perspective that we agree that climate change is indeed occurring and it is caused by human activity, and most of it (but not all) is caused by the energy industry. If we disagree on this critical point, I kindly suggest that you stop reading here, as I won't attempt to convince you otherwise. In fact, I'll be jumping back and forth between energy generation and climate change, because at least for me, the future of both is unquestionably linked (though other industries and human activities definitely contribute to climate change).

Instead, I want to discuss how and why we feel so detached from the energy issue. Unfortunately, this detachment is not exclusive to the general public. Many smart and well-educated people know very little about energy, and even hard-core environmental activists hold strong biases about energy generation.

I am always amazed to see our collective determination to keep energy in the political battleground. We all consume energy and therefore we should all care about making it clean, abundant, and affordable, regardless of our political views. We can support a fast move toward a renewable energy mix (or zero emissions) and remain skeptical about forceful governmental intervention and over-regulation. We could favour market-based solutions and still notice how energy markets continue to move in one direction, which promotes a massive fossil-fuel burning orgy (and not much else).

Watching political attempts to change energy policies can be very confusing and at times even amusing. For example, in 2014, the liberal Obama administration publicly discussed the idea of introducing Renewable Portfolio Standards[4], which is a legislative tool designed to mandate increased state-level production of renewable energy. The conservative reaction to this *idea* was a state legislation initiative given the ironic name of Electricity Freedom Act[5], which actually tried to protect the existing energy infrastructure and its 'right' to keep burning fossil fuels. You don't need to know much about energy policy or U.S. politics to feel confused. Liberals are trying to *force* more energy choices in the marketplace, while conservatives are trying to *protect* a non-competitive market – and both make terrible word choices. Energy can be so darn confusing for everyone involved.

So, we'll begin by examining what went wrong with how our society perceives energy across five dimensions, or what I call

---

[4] https://en.wikipedia.org/wiki/Renewable_portfolio_standard#United_States

[5] https://www.alec.org/model-policy/electricity-freedom-act/

collectively, the Energy Gaps.

The Energy Information Gap argues that the news and analysis we receive about energy is typically framed in negative terms, leaving us depressed and detached when energy issues are discussed.

The Energy Age Gap suggests that young people are rarely encouraged to work in the energy industry (and in some cases, are *actively discouraged* to pursue this field). This means that the future workforce and a new wave of innovators are still not in the game of reshaping the global energy landscape.

The Energy Innovation Gap argues that energy innovation has gone stagnant, and that energy innovations aren't emerging at the same pace that we're seeing in other fields.

The Energy Indecision Gap outlines how our governments have traded energy reformation for energy stability, leaving us with few energy options for the future.

The Energy Inspiration Gap explores how popular culture has left our energy future in the hands of the few, by making us all feel bored or scared to death whenever energy issues are raised.

The second part of this book, **The Make Energy Cool Blueprint,** proposes a new energy landscape. If you remember just one sentence from this book, to be shared at a party, it is this: an alternate, clean, and exciting energy future could begin right now, but only if we bring energy generation and energy consumption to the same *location*.

Spoiler alert: I won't suggest any specific new technologies to be used in this blueprint, but rather, I provide some economic, technological and cultural frameworks for how to make energy generation, not just its consumption, relevant to as many people as possible. In fact, one of the main foundations of this blueprint is that we don't need to invent much or make dramatic changes in our lives in order to make energy more relevant and accessible for everyone. In the proposed blueprint, energy would still work for us, and not the other way around.

This blueprint is presented from three angles:

Make Energy Markets Cool suggests that a new DIY or retail energy generation market could emerge, and it would have a profound impact on how we perceive and consume energy.

Make Energy Innovation Cool explores new ways to foster better and faster innovations in energy generation technology, rather than suggesting which current technologies offer the best hope for change.

Make Energy Inspiration Cool asks fellow creatives to start their own energy inspiration projects, and offers some insights from other fields that have undergone similar changes. This part also suggests how the new energy landscape can be used to generate an even deeper social change.

Finally, I've added an epilogue, which contains two parts:

Best Case Scenario attempts to explain how the future of energy could look if some of these ideas were implemented. I tried to be as specific as possible, though many technological, social, legal, environmental and economic issues remain unaddressed.

Worst Case Scenarios is the juicier part, where I challenge my own convictions in order to see whether my energy theories hold up, or if some (or all) of my original assumptions are wrong.

This book and its ideas are not perfect, nor are they are risk free. Any new technology or new system architecture has both foreseen and unforeseen implications. Our current global energy system works; it is powerful, stable and affordable. We know how to operate it, and we've kind of learned how to live with it.

But this system, this devil we know, is also broken in many ways, offering slow innovations, poor public image, a lack of widespread inspiration, and a terrible environmental impact. If you disagree with me on its environmental impacts, you can agree that energy production has a terrible public image. If you disagree with my thoughts about the energy sector's innovation gap, we can share a concern for how uninspiring energy production has become for younger generations.

A disclaimer to students: this book isn't a scholarly work designed to be quoted, nor is it supported by endless tables and documents. I tried to find as much data as possible to support my hypotheses, but I didn't conduct original research and, as such, there are no 'smoking gun' documents for every sentence you'll read. Please treat this book like a *blog post gone rogue*, but feel free to use and test its basic ideas.

Most importantly, I welcome all critics, ideas, thoughts, and suggestions. I've created a website at makecoolenergy.com and social media pages where we can continue this conversation. I sincerely want to hear what you have to say about this topic and how we can make energy cool. I'm looking forward to hearing *your thoughts* about our energy future.

Part I

What Went Wrong

or

The Energy Gaps

1.1 THE ENERGY INFORMATION GAP

Close your eyes and consider the most recent piece of information you've heard about energy or climate change. If you work in energy, or you're a climate/environmental activist and receive such information on a daily basis, try to think about the last media piece you've seen on these topics. I'm not talking about daily oil barrel or petrol prices, but rather, about an actual news piece featuring new developments in the energy sector or exploring climate change.

Regardless of where you live, this news coverage probably had three main characteristics. First, it was negative, either in terms of the information it conveyed or the images used to accompany it, such as smoking chimneys, endangered polar bears, or oil spills. Second, it was highly complicated, describing emerging issues or technologies that still felt secretive and complex, despite reporting intended for a general audience. And finally, it probably mentioned a gigantic energy company, or at least one or more governments were referenced in the coverage.

These three dimensions – negative coverage, secretive or unreadable information, and a focus on large, intimidating entities – has been the leading narrative of most mass-media energy coverage for the last 40 years, and probably much longer.

Since the early days of the Industrial Revolution, energy production has become essential to human activity, but it has always been controlled by few and considered a nuisance for everyone else. The very words "Industrial Revolution" throw us back into smoggy, Dickensian London streets, where pale-skinned orphans walked barefoot while inhaling the toxic fumes from thousands of coal-burning chimneys. Then there are more current smoggy scenes in places like modern-day China, and we still haven't mentioned Fukushima or Three Mile Island. In our collective minds, energy production seems to always be associated with the three Ds: Dirty, Dangerous and Dishonest.

When we look at how we're currently debating electric energy, one of the world's most desirable commodities, it's clear that energy and climate information is typically conveyed in a grim, politically complex and highly technical fashion. The technical complexity is often inevitable, because energy generation, like all other types of production, has become unbelievably complicated.

The grim discourse, however, is completely a matter of choice. More likely it's a deliberate choice made by energy PR experts and technical analysts. There is an incentive for large economic players with strong financial and geopolitical interests to make energy seem unbelievably complicated. This way, the regulators and the general public are too afraid of making necessary reforms or disruptive changes, because any change feels next to impossible and appears to demand a high price tag.

In a sense, the way we portray energy reflects how we cover all current affairs: things are bad, getting worse, and we have limited choices, because none of it is actually in our hands. In some ways, the grim portrayal of energy and climate issues makes sense. The current situation, frankly, is pretty grim. Bleak news also inspires more headlines and creates a sense of urgency.

Let's face it, this grim tone has been pretty successful and has even paved the road leading up to the COP 21 Paris agreement. Governments, NGOs, individuals, and even corporations fear dire circumstances and often work to minimize risks. It's human nature. But by now, this grim discourse is mostly a habit, rather than carefully crafted strategy.

Focusing energy stories on the interests of governments and large multinational corporations is also a matter of choice. In fact, many large corporations have gone out of their way to make (at least some parts) of their work more transparent.

In most Sustainability Reports, now released by all the major energy companies, corporations often share *their thoughts* about the future of energy, and their current efforts *to make it more sustainable.*

But even they can't be surprised that most of us are unaware or remain indifferent about their carefully written sustainability reports. While they try to sell us an optimistic story, we tend to see an underlying, darker narrative[6]. So, corporations don't inform us any more positively about energy issues than other players and platforms.

There are two other actors who can contribute knowledge to our collective energy perceptions. The first are, of course, our governments and regulators, who are supposedly the ones at the steering wheel. But if you think that governments are perceived as bearers of good news, then this is also becoming less common[7], because it requires them to provide information more willingly, openly and consistently –

---

[6] Thus, when British Petroleum (BP) worked to rebrand itself as *Beyond Petroleum*, its 2010 Deepwater Horizon oil spill in the Gulf of Mexico made us forget what lies Beyond Petroleum, and instead, focus on how bad it can be right now.

[7] See this piece about government transparency:
http://journalistsresource.org/studies/society/news-media/overcoming-negative-media-coverage-government-communication-matter

something that most governmental agencies find very difficult to do. The global sentiment, from Occupy Wall Street to the streets of Athens, Tel Aviv or Madrid, now seems to distrust most governments when they discuss *anything* that concerns our future. In the minds of so many people, governments are more likely to work for big business than for the little man or woman, and therefore, their information is also perceived as biased.

The last party that can help us get our energy facts and narrative right includes activists and Non-Governmental Organization (NGOs). While their resources are much more humble than those of energy giants, governments, and commercial media outlets, NGOs, too, tend to choose a dismal narrative, partially because it's easier to grab the media's attention when things go wrong, but mostly due to the fact that most activists believe that energy issues are, in fact, getting worse.

The result of this standoff – whereby media and activists tell us a grim story, while governments and corporations try to sell us a more positive one – is that we rarely get an alternative perspective, and only randomly hear positive energy developments from a source that we consider to be unbiased.

On the energy front, we seem unable to lose this negative attitude even when we turn away from traditional media outlets and seek out other sources.

In 2015, a study[8] conducted by scholars from four different countries comparing climate change reporting in mainstream media and millions of tweets during a period of 17 months, found that the negative, anti-energy sentiment has been replicated in the new media. The one common thread between both sources seems to be a reluctance to present climate issues in a positive way; less than 20% of all reports had a positive angle.

And the bad news continues for specific energy resources, most notably nuclear power – the energy sector equivalent of ex-convicts who everyone expects to make a lethal mistake, until the day when they finally do.

In 2012, notable journalism scholar Cristine Russell

---

[8] Olteanu et al, Comparing Events Coverage in Online News and Social Media: The Case of Climate, 2015 Change, http://www.aaai.org/ocs/index.php/ICWSM/ICWSM15/paper/view/10583

questioned[9] the media's ability to share an optimistic view of nuclear energy:

The question is whether the pessimistic tone of the [Fukushima] anniversary coverage was an echo of those mournful and angry images or something more organic. Pundits and the media can have short memories, and one day the techno-optimism might return. Only more time will tell whether improvements in nuclear safety and technology will help fuel a turnaround.

If you think that coverage was pessimistic because nuclear energy is unpopular, think again. A study by a British NGO[10], the Public Interest Research Centre, found that in 2011, most mainstream coverage of renewables in the British newspapers wasn't favourable (though a few papers have shown a more balanced approach). Prevailing attitudes could have changed since the release of this report, but at that particular point in time, even in the green-conscious U.K., it seems mainstream media hadn't quite decided if renewables were a good or a bad

---

9 Russell, Pessimism Reigns a Year After Fukushima, 2012, http://www.cjr.org/the_observatory/pessimism_reigns_a_year_after.php?page=all

10 PIRC, The portrayal of renewable energy in the media, 2011, http://publicinterest.org.uk/wp-content/uploads/2011/07/renewables_in_the_media.pdf

development.

Even noble attempts to escape the current affairs playground, like the game-changing 2006 film, *An Inconvenient Truth*, have adopted this grim tone. The lion's share of this important and highly influential project was spent on doom and gloom narratives, leaving the "what you can do about this" discussion for the end titles.

John Grisham's latest novel, *Gray Mountain* (2014), is another attempt to inform us about energy issues beyond the realm of current affairs. Grisham's noble effort, set in Appalachia in the wake of the 2008 financial meltdown, is a typical three-Ds scenario: America's coal land is described as a bleak hell on earth, ravaged by environmental and social degradation, and controlled by greedy and ruthless energy giants. Any regulatory intervention is usually corrupt and works in favour of deep-pocketed corporations. In this modern dystopian landscape, a few brave lawyers are all that stand between the earth and its poor, unfortunate residents, and this powerfully evil system. Sadly, there's not a single sentence about why we need all that coal and how we might be better off without it in Grisham's master medium – an engaging and highly readable legal thriller.

To put these news and energy information sources in perspective, if everything we learn about climate or energy is packaged in two-minute sound bytes or the occasional tweet, we're more likely to hear about lethal storms, oil spills, complex multinational agreement negotiations, or drowning polar bears. Most of these short clips seemingly have nothing to do with us, and even worse, they say next to nothing about how we can help to change the current situation.

Sadly, the only players committed to portraying our energy future in a somewhat positive light are the energy giants. If the future of energy were a spectator sport, there is currently just one team on the court. Even if that team were comprised of all-stars, we probably wouldn't buy a ticket or call it a match, and we wouldn't tune in for the game, because we already know how it's going to end.

1.2 THE ENERGY AGE GAP

On November 1, 2015, three weeks before the COP 21 convened in Paris, the Pew Research Centre published a survey of global public opinion[11] about climate change, revealing that the world population sent its leaders to Paris with a rather decisive backup; the global median of people who think that climate change is a very serious issue was 54%. Another 51% said that they think climate change is hurting people now, while 40% were concerned that they would be personally affected by climate change.

There were, of course, regional variations in the survey. The residents of Africa (61%) and South America (63%) felt the most vulnerable, while the Chinese respondents felt the most immune to the impacts; only 15% felt personally threatened by climate change.

An earlier poll by Pew, released in July 2015[12], revealed a more

---

[11] Stokes et al., Global Concern about Climate Change, Broad Support for Limiting Emissions, 2015 http://www.pewglobal.org/2015/11/05/global-concern-about-climate-change-broad-support-for-limiting-emissions/climate-change-report-29/

[12] Carle, Climate Change Seen as Top Global Threat, 2015, http://www.pewglobal.org/2015/07/14/climate-change-seen-as-top-global-threat/

interesting analysis. It tried to determine where climate change ranked in relation to other global or regional threats. In a survey that checked the public view across 40 nations and more than 45,000 people, climate change was *the most important global threat*, and people in 19 countries agreed that it was the world's biggest headache.

Nevertheless, there is a declining interest in climate change in rich countries, which has been observed for some time, like in a 2010 Gallup poll[13], or in this 2015 meta-study[14] conducted by several British researchers.

It seems that we are now experiencing a counter reaction to the last decade's heated climate change debate. The rich world is slowly losing interest, perhaps feeling more resilient, while residents of the developing world seem to feel that climate change has shifted from a scientific theory to a reality.

---

[13] Pugliese & Ray, Fewer Americans, Europeans View Global Warming as a Threat, 2011, http://www.gallup.com/poll/147203/fewer-americans-europeans-view-global-warming-threat.aspx

[14] Capstick et al., International Trends In Public Perceptions Of Climate Change Over The Past Quarter Century, 2015 http://onlinelibrary.wiley.com/doi/10.1002/wcc.321/pdf

A 2014 Gallup study[15] of American public opinion on the nation's most urgent issues ranked the *availability and affordability* of energy as number 10, right after *the possibility of terror attacks on the U.S. Climate change* reached number 15, beating only *race relations*, which, sadly, placed last on the 16-topic list.

A more fine-grained analysis of American opinion was made by Pew in 2015[16], when they tried to identify public opinion on climate change by demographics. While gender seemed to be insignificant, ethnic divisions were very important, as 70% of Hispanic Americans attributed human activity to climate change. That's 26% more than the 44% of white Americans who made the same link. Age was the other deciding factor. Most people (51%) younger than 49 years old were convinced that humans have contributed to climate change. Above the age of 50, the people holding this opinion became a minority, and above the age of 65, only 30% thought that climate change could be caused by human activity.

On a global scale, there is a similar trend, but with a much

[15] Riffkin, Climate Change Not a Top Worry in U.S., 2014,
    http://www.gallup.com/poll/167843/climate-change-not-top-worry.aspx

[16] Funk & Rainie, Chapter 2: Climate Change and Energy Issues, 2015,
    http://www.pewinternet.org/2015/07/01/chapter-2-climate-change-and-energy-issues/

more subtle nuance. In a 2014 survey[17] by the *British Chatham House,* nearly 85% of people under the age of 45 agreed or strongly agreed that human activity was changing the climate. Above this age, the percentage declined mildly, to 80% agreement.

So, seemingly the climate and energy fields are doing fine, and the awareness of climate change and the future of energy is on the right track.

The Millennials' Dilemma

On the other hand, opinions expressed by young people in surveys are only half of the story. The real question is what they're doing about those beliefs, and if and how they see themselves as responsible or involved. Here, too, the initial and theoretical verbal statements are highly positive. In the Deloitte Millennial survey of 2014[18], 65% of the millennial generation workforce (people born after 1980) thought that climate change was a responsibility of business.

---

[17] Carbon Brief, Insights from a global survey of climate change opinion, 2014, http://www.carbonbrief.org/insights-from-a-global-survey-of-climate-change-opinion

[18] Deloitte, Mind the gaps The 2015 Deloitte Millennial survey, 2015, https://www2.deloitte.com/content/dam/Deloitte/global/Documents/About-Deloitte/2014_MillennialSurvey_ExecutiveSummary_FINAL.pdf

But in this survey, we can start to see the growing cracks between best intentions and personal choices. While 78% of Millennials said that the company's overall innovation is a major consideration before committing to a workplace, 70% said that they would prefer, at some point in their careers, to work independently. And 61% thought that operational structures and procedures are an organizational barrier to fostering a culture of innovation, which they expect from their employers.

Needless to say, the current energy sector is hardly an ideal place for Millennials; it offers great stability, but limited opportunities to work independently. Working in energy can create a huge impact, but due to safety codes and regulations, reinventing the wheel on a daily basis is next to impossible, and the energy sector is not considered to be the most innovative environment to begin with. For example, not a single energy firm is included in the 2015 Forbes 100 most innovative companies list[19].

If you work in the energy sector and you don't think Millennials might be either frustrated by working in your

_____

[19] http://www.forbes.com/innovative-companies/list/#Services

organization or averse to joining the company in the first place, then it's time to test this idea empirically.

I couldn't find a reliable source of global statistics on the age of energy sector workers, but the US Department of Labor provides data on age structure and wages in various industries. Since the U.S. is the world's largest energy market, this information provides a glimpse into how attractive the energy sector can be – or sadly, how unattractive it has become.

We've already seen that the younger you are, the more likely you are to support the idea that humans have an impact on the climate. We've also seen that more young workers feel there is a problem with how business is addressing climate change. And finally, we've seen that Millennials and younger workers are not just looking for a job; they're seeking a sense of meaning, a cause. The combination of these three preconditions would be enough to send them flooding the energy industry with emails and CVs. Only so far, that hasn't happened.

An Aging Industry

The table below shows the 2013 median age, the age structure of workers split at 35, and the median income for several U.S.

industries, across both services and manufacturing: Energy, Architecture and Engineering, Moving Pictures, Computer Systems, Financial Services, and Automotive. (Absolute workforce numbers are in thousands).

Table 1: Age Structure of Selected Industries in the US, 2013

Industry	Total Jobs (000), 2013	Workforce Share < 35, 2013	Workforce Share > 35, 2013	Median Age, 2013	Median Annual Income ($000), 2014
Electric Power Generation, Transmission & Distribution	**595.0**	**25%**	**75%**	**46.0**	**$72.8**
Computer Systems Design & Related Services	2310.0	33%	67%	40.9	$87.9
Motion Pictures & Video	404.0	52%	48%	34.4	$56.5
Motor Vehicles & Motor Vehicle Equipment Manufacturing	1186.0	29%	71%	43.3	$53.8
Securities, Commodities, Funds, Trusts, & Other Financial Investments	1150.0	29%	71%	43.0	$101.0
Architectural, Engineering, & Related Services	1495.0	30%	70%	45.0	$74.6

Data source: US Department of Labor

For the "old" folks, or people born before 1980, the energy sector is doing fine, where it is providing an above-average median annual income (72.8K USD) versus the national median income of (47K USD). Wages are also competitive with potentially more esteemed occupations, such as Architecture and Engineering. Energy workers certainly fare better than those in the auto industry. Energy is also providing sought-after occupational stability, as it offers the most stable choice for people above the age of 45; nearly 52% of the industry's workers are over 45.

But even a competitive income doesn't score top points anymore, especially not with Millennials[20]. The median age of energy workers is 46, so it's the oldest industrial sector on this list. In the old world, we could expect that potentially lower wages and a risky future would deter newcomers from entering a fragile industry like Motion Pictures. Think again.

Making movies is so cool that Motion Pictures is the youngest industry on the list; 52% of film workers were born after 1980.

---

[20] Schulte, Millennials want a work-life balance. Their bosses just don't get why, 2015, https://www.washingtonpost.com/local/millennials-want-a-work-life-balance-their-bosses-just-dont-get-why/2015/05/05/1859369e-f376-11e4-84a6-6d7c67c50db0_story.html

In comparison, only 25% of the energy workers were born during this period.

In the graph below, I combined the energy industry age structure, the general population age distribution and the Pew survey results, by age. While we don't know what lies in the hearts and minds of the people who currently work in the energy sector, the American population slice with the highest confidence in climate change science, the younger generation, is the least represented age group in the industry.

Graph 1: Share of Support for Climate Science, Age Distribution in Energy Sector & General Population Age Structure , 2013

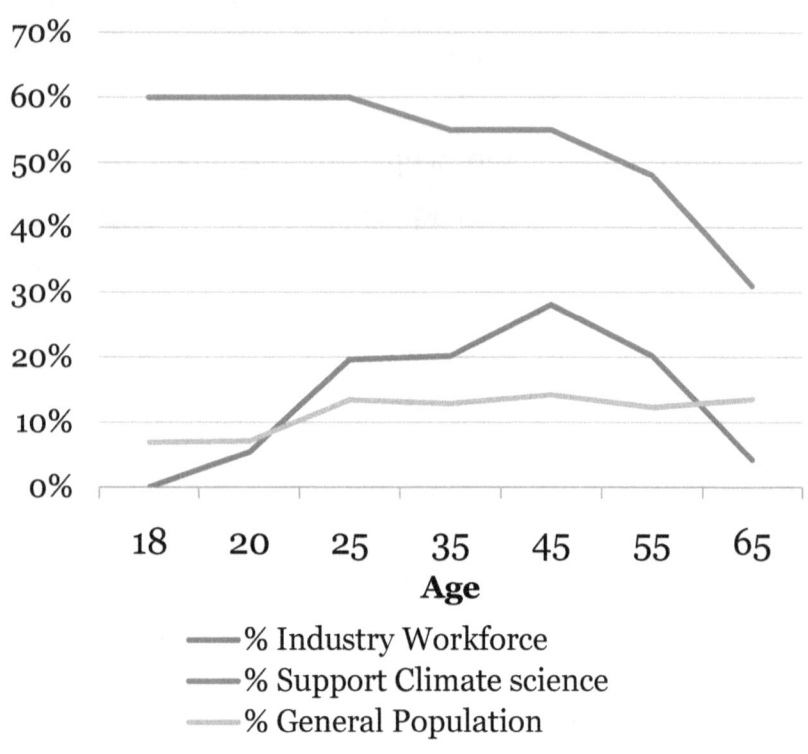

Data source: Pew, US Department of Labor, US Bureau of Statistics

If you think that I've compared apples to oranges, or that there are other ways to look at this age gap, then let's take another approach. In 2013, Joshua Wright of *Forbes* magazine conducted an analysis[21] of key skills gaps across major U.S. industries. His findings showed that the oldest skilled profession in America, with more than *70% of skilled workers above the age of 45*, was *Electrical and Electronic Repairers*, followed by Extruding and Drawing Machine Setters, *Electrical and Electronic Engineering Technicians, Stationary Engineers and Boiler Operators, Maintenance Workers in Machinery, Electricians*, and *Computer Controlled Machine Operators* (professions in italics are all energy-related), which represented the first group of the most rapidly aging skills in America.

This age gap is not just an American problem. It's happening elsewhere. I did my bachelor's degree in the Faculty of Architecture at the Technion, Israel's Institute of Technology. The neighbouring faculty was Electrical Engineering.

I've translated some of their landing page text in order to demonstrate how a department, which claims to be one of the 10 most important electrical engineering faculties in the world, is trying to attract new students:

---

[21] Wright, America's Skilled Trades Dilemma: Shortages Loom As Most-In-Demand Group Of Workers Ages, 2013, http://www.forbes.com/sites/emsi/2013/03/07/americas-skilled-trades-dilemma-shortages-loom-as-most-in-demand-group-of-workers-ages/#2715e4857a0b7711dc674545

"The Electrical Engineering Faculty is involved in a multitude of high-tech professions, including communication and information, voice and image processing, computers, computer and data networks, multimedia, software engineering, electro-optics and optic communication, electromagnetic waves and fields, micro and nano-electronics, VLSI, control and automation, bionic systems, medical electronics and bio-transmitters, visual and photographic science, nerve-networks and self-learning systems"[22].

I could have saved you from reading this list of fascinating research and professional disciplines, because the essence is this: if you're going to be an electrical engineer, you can work in any field, except the fields of *energy generation or energy transmission.*

This faculty isn't trying to hide a wonderful profession from the rest of the world. They're just going with the flow. All of the disciplines in this particular faculty's list are now more popular than designing the next power plant or improving the national power grid, and the young students seem to believe

---

[22] Don't bother looking for this text on their English website <u>here</u>, as it is not intended for the local demographics.

that they will have better and more rewarding career options in other fields.

The Soldier-less Generals

Deloitte's survey clearly demonstrates that Millennials are looking for a sense of purpose in their professional lives. The energy sector can provide this purpose. But in its current state, it seems unable to attract young minds. This doesn't mean, however, that people who are 35, 58 or 75 can't come up with new and exciting ideas – they most certainly can.

It also doesn't mean that energy sector progress and innovation can't come from the outside, such as through private R&D efforts or universities[23]. It can. But in a hyper-connected world, where knowledge, social communication, and creativity are the true means of production, young newcomers are a crucial part of reshaping the future energy landscape, and when their numbers are low, their potential contribution decreases proportionally.

These young professionals, too, could benefit from being more involved in the energy sector. They could gain more knowledge

---

[23] Though, as we'll see in the next chapter, energy innovations rarely come from these outsiders.

about the current industry's flaws and damages on one hand, and of its importance on the other. But they need to be insiders in order to gain these insights, versus reading about it in books or on Twitter.

This new workforce is smarter, better educated, more innovative and more self-motivated than any group we've known in the past. They despise stability and rarely consider compensation to be the deciding factor in their professional choices.

But they're also quick thinkers and learners, loyal to self and wanting to see fast results.

Could it be that the information they receive about this industry – shared through books, the old and new media, and the films they watch – tells young, creative workers that energy is just not a good fit for them? That it's too old, too dirty, too dangerous, and too dishonest?

Is energy becoming *No Country for A Young Man (or Woman)*?

Dismissing the age gap as a function of supply and demand is too easy. I don't think that this is an employment or a wages problem. The real issue is that we are quickly approaching what is probably the most historically important period of energy innovation, which could also be our last chance to prevent an unthinkable and irreversible climate change, but we're running ahead with fewer numbers and breathing a little heavier with the effort. Our senior energy engineers and executives may have secured their retirements, they may be our best men and women to lead the charge, but they are quickly becoming *soldier-less generals.*

1.3 THE ENERGY INNOVATION GAP

The word *innovation* is often presented as an ultimate measurement of national and industrial progress. Politicians and companies alike love to boast about how they *create and promote* innovation. Innovation sounds promising and impressive, but it is actually quite difficult to quantify and measure. Throwing the word "innovation" around is a bit like wearing chic glasses as an accessory; they don't hurt your appearance (at least when you're older than 12) and they usually make you look smarter.

While there are several possible indices, innovation is often measured through one of two approaches. The first way measures investment in innovation, which refers to how much effort and how many resources are dedicated to innovation research and development (R&D) in a company, industry, or in a political entity such as a state, city or region. We will discuss energy R&D investments in another chapter, **The Energy Decision Gap.**

The second approach measures innovation outputs, such as the number of new patents, the number of available products and services in a sector, and more. But this measurement is hardly straightforward; innovation often occurs in incremental ways, like improving automotive fuel efficiency over many years (evolutionary). It can also occur at a much faster pace (revolutionary), such as Tesla introducing an all electric-vehicle that suddenly makes fuel efficiency advancements seem obsolete.

In this chapter, I have deliberately preferred to measure energy innovation by revolutionary standards rather than by evolutionary progress. While evolutionary innovation is an important part of any industry, it's the revolutionary innovation, the huge leaps forward, that make our jaws drop and launch endless possibilities and further innovations. Think about the first time you saw a mobile phone with downloadable apps, the first affordable drone, or the first consumer 3-D printer. All of these products led to further waves of innovation.

If we look at energy innovation, these huge leaps didn't occur for a very, very long time. Some of you might think this is an unsound statement, because we keep hearing stories about new technologies, first-of-their-kind projects and exciting energy developments, typically in the field of renewable energy. You may have even seen variations of this graph, which shows the sharp rise in renewable energy[24] output (in TWh, or Billions of kWh) over the last three decades:

Graph 2: Global Electricity Output, Renewable Sources (in TWh)

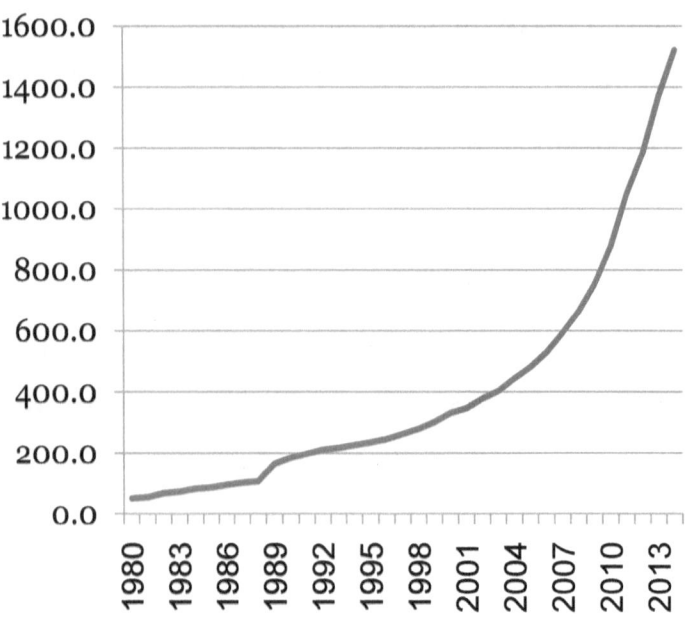

Data Source: The Shift Project[25]

---

[24] Incl. the following sources - Geothermal, Wind, Solar, Tide and Wave, Biomass & Waste, and Hydroelectric Pumped Storage.

[25] This fantastic, free and thorough database, the Shift Project, is ideal for anyone who wants to understand the trends and projections of world energy production. I found their work to be quite reliable and thorough.

When we look at this graph, it seems like renewable energy is the perfect investment. For example, if you bought $100 worth of renewable energy shares back in 2000, the 2014 value of that investment would have been $618. It looks pretty impressive when we compare it to this graph, which shows the same measurement for fossil fuels[26]:

Graph 3: Global Electricity Output, Fossil Fuels (in TWh)

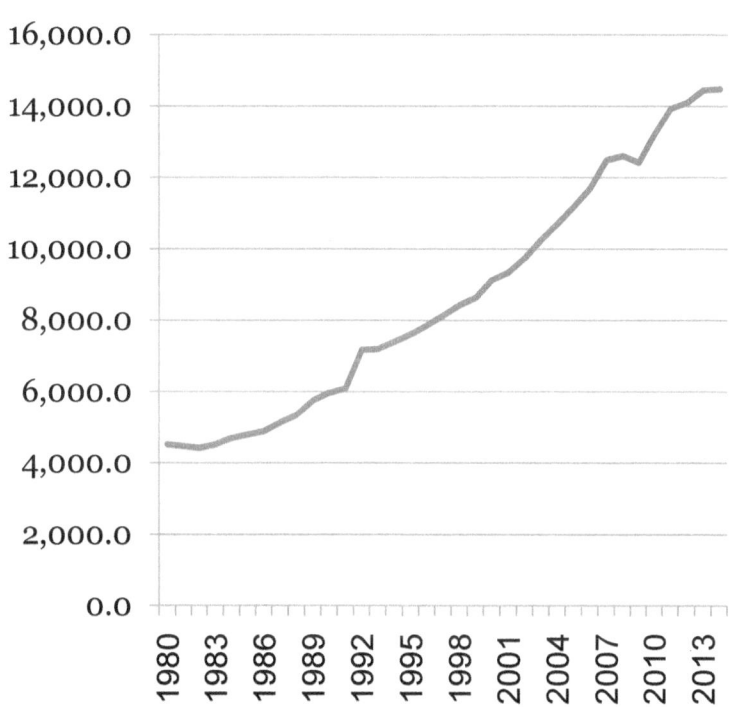

Data Source: The Shift Project

_____

[26] Incl. the following sources - Oil, Coal and Natural Gas.

This graph shows that an equivalent investment in this segment yielded 'only' $141, so we were much better off placing our money on renewables.

Assessing the last 15 years in energy innovation might be a bit unfair, given the time it takes to build new infrastructure (in most places, sizeable infrastructure projects take about 10 years from early concept to full capacity operation). So, let's double our time travel and go *Back to The Future*, to 30 years ago. In 1985, when Doc Brown and Marty McFly taught filmgoers that 2015 would have flying cars, hover boards and self-lacing shoes, they were wrong. Sadly, they were also wrong when they thought that the Delorean could run on the nuclear fusion of a banana peel and leftover beer.

In order to understand what would happen to Doc, Einy and Marty when they flew over our 2015 power plants, I've organized key **Shift Project** energy figures from 1985 and 2014 (the last year of confirmed data) in the table below[27].

---

[27] 'New' Renewables include the following sources - Geothermal, Wind, Solar, Tide and Wave, Biomass & Waste, and Hydroelectric Pumped Storage; Fossil include the following sources - Oil, Coal and Natural Gas.

Table 2: Global Electric Energy Mix - 1985 vs. 2014

	Output 1985 (TWh)	Share 1985 (%)	Output 2014 (TWh)	Share 2014 (%)	Change in Output (%)	Share Trend (%)
Total 'New' Renewables	75	1%	1,498	7%	1997%	6%
Total Hydro-power	1951	23%	3,769	17%	193%	-6%
Total Nuclear	1425	17%	2,417	11%	170%	-6%
Total Fossil	4928	59%	14,727	66%	299%	7%
Total Global Output	8380	100%	22,411	100%	267%	-

Data Source: the Shift Project[28]

It's time for a little bit of historical context. I chose 1985 based on the time that has elapsed (30 years), but it also marks the end of one of the most historically important eras of energy innovation. In the wake of the 1970s Oil Embargo[29], governments worldwide had no choice but to pursue alternatives to crude oil.

-----

[28] I should note at this stage that some of you might be familiar with much higher renewables figures from other information sources: REN21 estimates for 2013 are of 19% renewables; the International Energy Agency figures for renewables' share in 2013 were 22%; and the European Commission's figures for its 28 members was 24.3% of energy from renewables.

I certainly don't accuse these sources of any manipulation, secrecy or bias. They are well-funded, well-respected organizations with their own methods and professional work standards. At the same time, these data sources are limited either in the time periods they cover (most don't show historical data for more than 10-15 years) or some, like the IEA, require payment to see raw data. But if you've read this far, you've probably noticed that I think that energy data should be free, comprehensible and accessible to all. I think the Shift Project fits this standard, and I prefer working with an open database.

There are also some nuanced differences between these information sources and what is being discussed here. For example, when the IEA covers renewables, they also talk about energy for transportation, while this book focuses on energy generation for electricity. Another major difference is that these sources usually count hydropower as a renewable, while the real purpose of this chapter is to analyse innovation in energy, not just resource diversification. A revolutionary innovation needs to be relatively new, but also, to make a big impact over a short period of time. It needs to be disruptive. I don't think hydropower quite meets this criterion. Hydro has been commercially used for more than 100 years. It isn't a totally new concept introduced on a large scale in the last 30-50 years, like nuclear, solar and wind, and it is now actually in decline, so it didn't change the rules.

[29] https://en.wikipedia.org/wiki/1973_oil_crisis

The result was the first wave of energy resource diversification, and until this very day, the most important. Some of the best renewable energy breakthroughs have occurred during the '70s. But these were marginal in comparison to the race for nuclear energy.

As we can see in the graph below, in the decade between 1975 and 1985, global nuclear energy output jumped from around 400 TWh per year to 1,950 TWh.(487% increase). In the following decade, the global output increased to only 2,210 TWh (113% increase, less than a quarter of the previous rate). Nuclear energy growth has continued to slow, and it is now actually in decline, as very few new facilities are being commissioned.

Graph 3: Global Electricity Output, Nuclear (TWh)

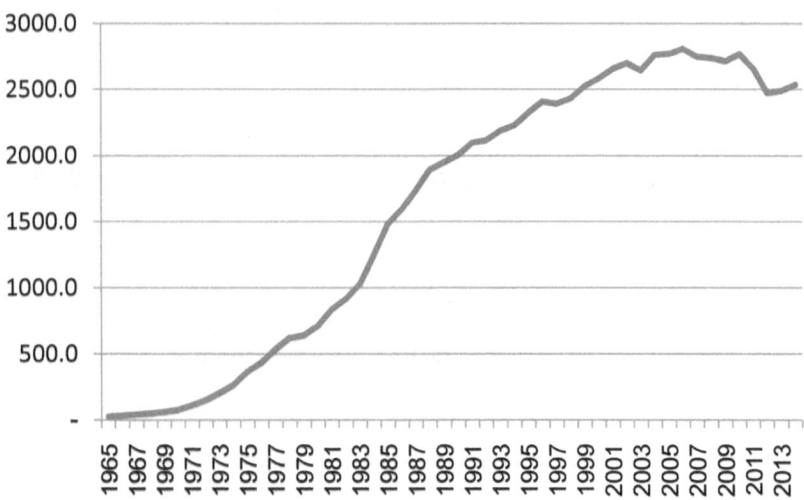

Source: BP Data Workbook[30]

Between 1985 and 2014, the growth rate of "new renewables" (all renewables, excluding hydropower) was 1,995%, which can look really impressive. That is, until you look at two other figures. The first is the *new renewables'* share of the global output, which is 6%. In a business-as-usual scenario, this means that to be on par with fossils, (e.g. representing 50% of global output), new renewables would have to wait until one sunny day in 2041, so don't hold your fossil-fuels-polluted

---

[30] http://www.bp.com/en/global/corporate/energy-economics/statistical-review-of-world-energy/downloads.html

breath[31].

The second, and probably even more discouraging figure, is the share of fossil fuels from the global energy output, which has grown over the last 30 years from 59% to 66%, literally increasing its share at 1% faster than the *new renewables* (and thus offsetting most of their progress).

If we somehow imagine that new renewables are compensating for the decline in 'dirty' nuclear energy, then fossils have increased their share at the expense of the 'clean' hydropower. If this were the music business, The New Renewables would now be a respected indie act. The Fossil Fuels Band would play the Super Bowl halftime show every year, for 30 years in a row.

So despite everything we hear, and how much our governments and energy giants talk about and spend on renewables, the global scorecard on this source is pretty lame. If new renewables were a private enterprise, designed to stop global addiction to fossil fuels, it would probably shut down after 10 years for spending too much money and making too

---

[31] Some might consider this 27-year turnaround an achievement. When we look back on our record over the last 30 years, and how easy it is to get distracted during this effort, I don't think that setting an alarm clock for 27 years from now and seeing how we do is a risk-free strategy.

little progress.

Now let's compare the last 30 years of energy innovation to innovation in other industries during the same period. If energy were Information Technology (IT), you probably wouldn't be able to read this book on your latest computer, because it would look like this:

Source: Wikipedia[32]

[32] Commodore 128, a popular personal computer introduced that year in Las Vegas
https://en.wikipedia.org/wiki/Commodore_128

The good news is that because we've doubled the output (in energy, we moved from 8K TWh in 1985 to 22K Twh in 2014), you could now have two of these lovely machines, or just one, with the *breathtaking processing power* of 256 bytes, which could process about one-third of this digital manuscript.

If energy technology paralleled automobile technology, and you wanted to purchase a 2016 vehicle, you might still see something like this at your nearest dealership:

Source: Wikipedia[33]

[33] The most popular vehicle in North America at the time- The Ford Escort
https://en.wikipedia.org/wiki/Ford_Escort_(North_America)

I can go on and on with these comparisons, but you get the point. Energy technology has literally stood still over the past 30 years.

The House Always Wins

There are other ways to look at it. Innovation theorists often talk about an *innovation ecosystem*, which is an entire regulatory, social, financial and cultural environment that promotes the creation of new ideas, inventions and enterprises. Let's examine how the current ecosystem is supposed to foster energy innovation.

Most R&D investments in energy and renewable energy are made either by governments or energy giants. According to this Bloomberg/UNEP report[34], governments and energy companies spent $12 billion out of $15 billion expended globally on renewables in 2014. The other $3 billion belonged to venture capital and private equity[35], which is low for new players in large industries. According to a 2013 study by

[34] Global Trends in Renewable Energy Investments, 2015, http://fs-unep-centre.org/sites/default/files/attachments/unep_gtr_data_file_11_may_201 5_amc_lm.pdf

[35] If this sounds like a lot, then you'll be interested to know that according to a strategy & consulting firm, it is less than what Apple alone spent on R&D during that same year ($4.5 BN).

Battelle, the world's largest nonprofit research and development organization, the energy industry has the highest rate of in-house innovation (75%)[36] among five major global industries: Life Sciences, Aerospace, Defense and Security, Information and Communications Technologies (ICT), and Chemicals and Advanced Materials. In other words, as an industry, energy depends mostly on its own resources to fund innovation and new ideas.

This self-reliance figure would be promising, but the same report found that the energy industry also had, on average, the lowest rate of R&D intensity (less than 1%)[37], which is defined as investments in R&D divided by sales. In other words, as an industry, energy uses significantly less of its revenues on R&D than other comparable major industries.

If you think that $15 billion is still a nice, healthy figure that can generate real innovation, look at IT and Software R&D

---

[36] 2014 Global R&D Funding Forecast, 2013, http://www.battelle.org/docs/tpp/2014_global_rd_funding_forecast.pdf?sfvrsn=4

[37] Compare this figure to 10-25% R&D intensity in Life Sciences; 7-20% in IT and Software, 4-6% in Aviation and Defense; or 2-7% R&D intensity in the Chemicals Industry.

spending for the same year: $307 billion (2,600% more)[38].

Yet comparing energy and IT is a bit unfair, because computers represent a much bigger industry. But computers have something else that energy doesn't have, namely the ability to start a new enterprise and quickly make it big.

It's currently almost impossible to create a new energy development and market it directly to millions of potential clients, in the same way you can with cars, software, chemicals, drugs or consumer electronics. You can't build a new power plant and immediately sell energy to consumers, because you need permits and licenses and you need to go *downstream,* or obtain access to the grid. By access to the grid, I mean two-way access: we can all consume energy, but very few of us can actually feed energy to the grid and trade it with other consumers. So instead of building new enterprises, entrepreneurs and investors need to *sell* their newest inventions and business models to those at the system's top – energy giants and/or regulators. This, in turn, creates a closed B2B[39] or B2G[40] innovation system, which can be referred to as

---

[38] This gap is actually widening. In 2014, investments in IT R&D were rising by 16% per year, while investments in energy R&D were rising by only 4%.

[39] Business-to-Business sales model.

Toll Booth regulation or Regulation of Entry[41], because it is so closely monitored by regulators. Breaking the rules, creating a disruption or breaking out of the pack is next to impossible in such an ecosystem. You can only invent something that'll *fit into the system*, not something that *creates an alternate system*.

So on one hand we have energy companies, which make a lot of money, spend too little on R&D, and usually prefer their own innovation; on the other, we have newcomers who need to sell their innovations to these existing incumbents.

Thus, we're subjected to an ecosystem that promotes incremental, discrete, *evolutionary innovation,* rather than the disruptive, *revolutionary innovation* we've come to know in other fields.

The biggest energy sector innovation from the last half-century, nuclear energy, had been mostly stagnant over the last 20 years, then finally rolled back. The hope of renewables has been diminished to a small percentage of total output. Our

---

[40] Business-to-Government sales model.

[41] Djankov et al, The Regulation of Entry, 2002,
      http://scholar.harvard.edu/files/shleifer/files/reg_entry.pdf

current energy mix looks suspiciously similar to what we had 30 years ago. Last but not least, the energy sector's carbon footprint has remained relatively steady[42] for the last 30 years, which some might consider an achievement. But with energy representing over 60% of Global Anthropogenic Greenhouse gases (GHG) emissions, we're nowhere close to taming this beast.

---

[42] International Energy Agency, Energy Climate and Change World Energy Outlook Special Report, 2015, https://www.iea.org/publications/freepublications/publication/WEO2015SpecialReportonEnergyandClimateChange.pdf

1.4 THE ENERGY INDECISION GAP

Why is the energy sector so slow to test and adopt new technologies? How did we lose 30 years of innovation in this sector?

There are a few explanations for this lag, but strong government involvement in the energy sector is the most crucial factor. The 1970s' oil embargo created global concern over the implications of using energy as leverage in global disputes[43]. Tensions caused by the 1970s crisis also sent energy regulators around the globe to pursue two often-competing agenda items. The first item was energy stability (which now, together with safety, is commonly referred to as Energy Security).

In order to ensure stability, governments have moved to tighten their grip on energy production and distribution. While historically, energy utilities in most places were typically state monopolies, even in countries where energy has been largely privatized, like the U.K. and the U.S., energy markets are always centrally controlled, planned and surrounded with

---

[43] This is not an unsound assumption and can be seen in several international crises, the most recent being the tension between Europe and Russia over Ukraine.

heavy regulatory and legal barriers, which prevent energy markets from running free (Regulation of Entry or Toll Booth Regulation). The official explanation for this control, as typically presented by politicians, is to ensure a reliable and affordable energy supply and to limit monopolistic abuse by the energy sector.

The regulators' second agenda item was to find alternative energy. Rich world governments invested heavily in this field during the late 1970s and early 1980s. But in the second half of the 1980s, oil prices went down, and suddenly, looking for alternative energy didn't seem like the best use of taxpayers' money. Like all public spending over the last two decades, alternative energy has moved towards a Public Private Partnership (PPP) model, where private money now plays a greater role. As a result, the last decade saw much greater investments in energy R&D[44]. According to the Bloomberg/UNEP report[45] we discussed in the previous chapter, in 2014, like most of the previous decade, the biggest

_____

[44] Kilisek, Higher R&D Investment in Renewable Energy Technologies Critical for Clean-Energy Innovation & Climate Action, 2015, http://breakingenergy.com/2015/05/13/higher-rd-investment-in-renewable-energy-technologies-critical-for-clean-energy-innovation-climate-action/

[45] Global Trends In Renewable Energy Investment, 2015, http://fs-unep-centre.org/sites/default/files/attachments/unep_fs_globaltrends2015_chart pack.pdf#overlay-context=publications/global-trends-renewable-energy-investment-2015

energy R&D spenders were private investors, representing $6.6 billion, or 56% of total expenditure.

A Kodak Moment

But if *private money* sounds like an early version of crowdfunding, or the more established venture capitalism, you're wrong. Usually the investors in energy R&D were the energy giants themselves. In 2014, like in most previous years, only $1 billion USD[46], or 7%, of renewables R&D expenditure came from VC, and the rest was from government or internal corporate R&D. The most notable example of internal corporate investment was, of course, BP, which re-branded itself as *Beyond Petroleum*. In 2010, the year of the Gulf of Mexico Deepwater Horizon incident, BP spent almost $780 million per year on R&D[47]. Let's stay on this figure for a minute. According to Bloomberg/UNEP, in 2010 the whole world spent $8.9 billion USD on renewables R&D. This means that about 9% of the global investment was made by a single firm!

---

[46] Global Trends In Renewable Energy Investment, 2015, http://fs-unep-centre.org/sites/default/files/attachments/unep_fs_globaltrends2015_chart pack.pdf#overlay-context=publications/global-trends-renewable-energy-investment-2015

[47] BP's expenditure on research and development from 2010 to 2014 (in million U.S. dollars), http://www.statista.com/statistics/302538/expenditure-on-research-and-development-of-bp/

In the year after the Gulf of Mexico incident, things got so bad for BP that it rolled back many of its investments to $636 million per year – an 18% decrease (it hasn't reached similar levels ever since).

In 2011, BP's sustainability report[48] said:

"We are working to become a simpler business, with a clear focus on what we do best. Our distinctive capabilities include exploration, operations in deep water, the managing of giant fields and gas value chains, and our world-class downstream business – underpinned by technology and relationships."

Taking this out of context and blaming BP for investing too little in alternative energy (they spent a fortune) or ignoring the problems associated with fossil fuels (they didn't) would be inaccurate and unfair. In fact, they've played a very important role in this effort. So important, that even as they were going down in the Deepwater Horizon's whirlpool of despair, they kept imploring the world's regulators to take a stand.

----

[48] You can read the full report here,
http://www.bp.com/content/dam/bp/pdf/sustainability/group-reports/bp_sustainability_review_2011.pdf

The same BP report included this paragraph (emphasis is mine):

"While we feel a strong responsibility to help meet this [fossil fuels] *growing demand, we also share widespread concerns about the rising global CO2 emission levels that it implies.* **BP supports government action to limit emissions and deliver a sustainable energy mix,** *including placing a price on carbon, increasing energy efficiency and providing transitional incentives that enable renewable energy to become competitive at scale."*

If our governments started cutting their spending on energy R&D and stopped believing in energy alternatives, why should the giants (who have the most to lose from investing) volunteer to support it instead? It's like expecting the managers of Hilton to spend their money developing Airbnb. Or expecting Kodak to unleash the potential for digital photography[49].

The competing agenda items facing energy regulators – to make the energy system stable while looking for its alternatives – became an impossible puzzle to solve.

---

[49] Pahal, How Kodak Squandered Every Single Digital Opportunity It Had, 2012, http://mashable.com/2012/01/20/kodak-digital-missteps/#tG4HVZOolZq4

Developing renewable energy while providing conventional energy is not like introducing a new flavour of Frito Lay potato chips or designing a new car at Chevrolet. It requires a totally different business model, technology, operation and mindset.

It's not surprising that regulators and energy giants have chosen to place their bets on the safety of fossil fuels. Following the Pareto Principle[50], which states that roughly 80% of all effects come from 20% of the causes, renewables took too much of their time, but generated too little revenue (for companies) and too little energy (for regulators). If they had to make a choice between the two, stability was the no-brainer.

Earlier, I wrote that there are no 'smoking guns' for every sentence in this book. Sadly, I couldn't find record of any meeting between regulators, energy executives, and lobbyists who sat together and decided to officially kill renewable energy[51]. It's likely that such meetings never really occurred, but that doesn't mean that a definite, but quiet killing didn't happen.

---

[50] https://en.wikipedia.org/wiki/Pareto_principle

[51] Like some have suggested was the case for the electric car.

Killing It Quietly

The graph below was published in a report[52] by the American Department of Energy, examining 57 years of government investment in energy R&D. While it might seem relevant to our investigation to see what happened only in the field of renewables, minimizing our analysis to only this investment program actually distracts us from the bigger picture. We can see that while there were various peak periods (early 1960s, mid 1970s, and early 1990s), the American government involvement has been very *stable* over nearly 60 years of record. While all of these peaks are associated with geopolitical issues (1963 Cuban Missile Crisis, 1973 Israeli-Arab War, 1991 Iraqi Invasion of Kuwait) they also distract us from the broader context, and that is the *overall stability* of government involvement.

---

52 Dooley, U.S. Federal Investments in Energy R&D: 1961-2008
    http://www.wired.com/images_blogs/wiredscience/2009/08/federal-investment-in-energy-rd-2008.pdf

Graph 4: US Government Expenditure on Energy R&D, 1961-2008

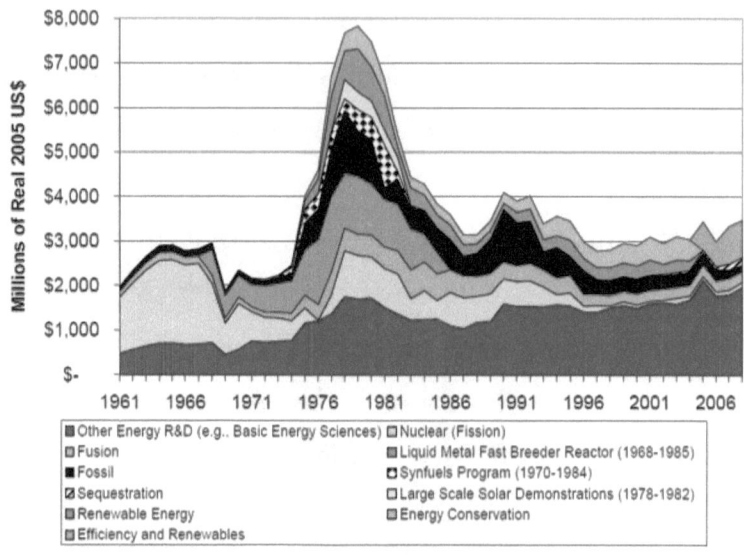

Source: Dooley, U.S. Federal Investments in Energy R&D: 1961-2008

If we take away the 1970s peak, then American government spending on energy R&D, until very recently the most important factor in global energy innovation efforts, has remained stable at $2.5 to $3 billion USD per year, for nearly 60 years.

The 1970s peak was not only the most dramatic period; it was also supposed to be the big game-changer. Geopolitics and oil prices were significant influences, but this period also represents a serious attempt to diversify our energy resources – a push made evident by the sheer number of investment streams. There were nine distinct R&D investment programs from 1973-1986, versus four in the 1960s, six in the 1990s, and there have only been five R&D investment programs since 2000. While these may include dozens of sub-topics and hundreds of different projects, we can see how unusual and bold the 1970s wave really was.

The consultants commissioned by the *Department of Energy* to write this report chose to mention this quote from the Reagan Administration in order to explain the 50% plummet in federal R&D expenditure that occurred in the late 1980s: *"only in areas where these market forces are not likely to bring about desirable new energy technologies and practices within a reasonable amount of time is there a potential need for federal involvement."*

Regulators, in the fashionable 1980s' *Reagan-Thatcher* laissez faire[53] – where *government intervention* in markets was labelled as *public enemy number one* – assumed that if something wasn't broken, there was no reason to fix it. Since the energy crisis was over (e.g. oil prices went down again), there was no need to publicly support energy innovation.

---

[53] https://en.wikipedia.org/wiki/Laissez-faire

History's Ironic Smile

Though I couldn't find a similar historic comparison on a global scale, the Organisation for Economic Cooperation and Development (OECD) has kept a record of some of its members' efforts in this field between 1974 and 2014. The graph below presents the OECD member states' investments in energy R&D, excluding the U.S. data discussed above (though data is still missing for some countries).

Graph 5: OECD (Excl. US) Expenditure on Energy R&D & Crude Oil Prices 1974-2014

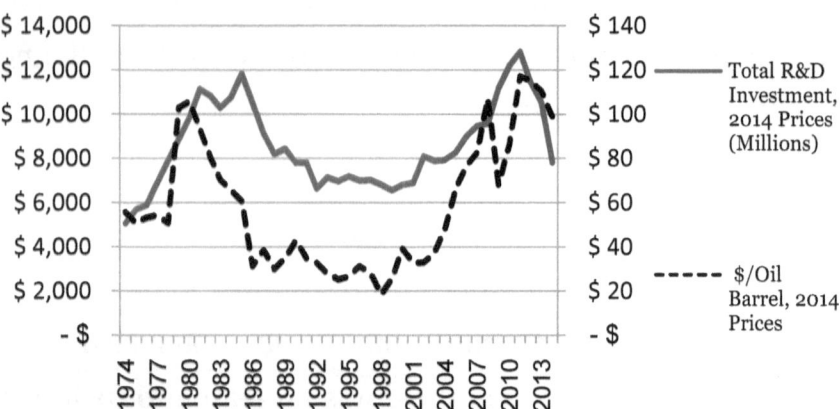

Source: OECD Library

We can see that even without Uncle Sam, global investments in energy R&D declined sharply between 1985 and 2000, in a graph that looks like history's ironic smile. Taking a snapshot of energy R&D investment policy in a rich world country in 1980 and another in 2010 may look promising, because we actually saw an optimistic wave of investment during those years. But in 2016, we already know how at these previous waves ended.

New investments started increasing again only when oil prices started to rise sharply during the second half of the last decade. These increased efforts were partially supported and explained by more robust scientific evidence of a man-induced climate change, which began to emerge in the early 2000s. Now, when oil prices go down again, these investments, too, seem to some as irrelevant, so it seems the climate story doesn't really stick.

On a six decades timeline, we can see that rising oil prices were the only real incentive for governments to invest in energy R&D. The COP 21 is again putting *all the power* to adapt and mitigate climate change back into governmental hands.

But while we have some influence on our governments' policies, as individuals and non-organized citizens, we have a very limited influence on global oil prices. We can trust our government to do the right thing, but we need to do more. We need to help our governments and the people working for them.

To be clear, I'm not suggesting that governments don't care or won't work to resolve climate change. They'll play a vital part in its resolution. But at the same time, placing all of our hopes and faith in entities that have different interests and competing agendas seems too risky, given what's currently at stake – and that is irreversible climate change. We can do better.

I have seen the work of government officials in several nations and in several levels of seniority. Contrary to common belief, these officials are not all corrupt, stupid, lazy or apathetic. There are some highly talented and remarkably motivated individuals in every part of the public sector, and from what I've seen, these talents are scattered in many places around the globe. But when a system works so hard to keep the status quo, those individuals are too weak to fight it.

Thirty years after the last global effort to reshape the future of energy, we are stuck with the same energy system, the same energy mix, and the same regulatory control. The only difference is that now we've wasted 30 years and spent billions on energy

sources that provide a few meager percentage points of the global output. Thirty years of indecision.

1.5 THE ENERGY INSPIRATION GAP

In a previous chapter, I discussed how misinformed we've become in relation to energy news. In this chapter, I'd like to focus on inspiration. There's a big difference between being informed and becoming inspired. We can be informed by many things, including rumours, media coverage, and social networks. But being inspired is much more complicated and often intrinsic. It requires an idea or information to reach deep within and stir us into action.

To make my point, I'll focus on blockbuster Hollywood movies. Incidentally, I didn't choose movies because I think they're the only medium that inspires us. We can be inspired by books, comics, games, radio, art, and many other forms of visual and verbal expression. But movies work in a fantastic way, inspiring us while we enjoy their artistic and visually compelling aesthetics and narratives. Blockbuster movies are also universally accessible (unlike literature, which requires a common language and literacy) and people from many different cultures and ages can enjoy their basic storylines. And finally, I've chosen them because blockbuster films are the ultimate media format survivors; they've outlived the TV, videocassette, DVD, VOD and quite possibly, the Blu-ray. In fact, they've even survived the biggest, most sophisticated and most badass competition of all, which is web piracy. It looks like movies are here to stay (lucky us) and they are a fascinating, ever-evolving medium. In short, blockbuster films are cool, durable and offer a profitable creative industry (clear evidence that such an industry can exist).

Generally speaking, the cumulative impact of the movies we watch changes how we perceive reality. But the best ones make us act upon that reality. This is probably the ultimate achievement of any creative project; to inspire and create positive change (all while selling millions of copies, of course). Yes, movies (like books) can inform us, but they also push us to learn new things and they inspire us too, without noticing it ourselves. Sometimes that inspiration can be good. Sometimes it's really bad.

Good examples inspire us to be bold, learn languages, and see what others are doing, creating and thinking. Bad examples, on the other hand, can inspire violence and hyper-consumerism, and (potentially) represent a global, lowest-common-denominator vision of creativity.

What's truly fascinating about blockbuster movies, though, is that they tend to help us process some concepts faster than others, including technological and political changes – but even here there are variations. If we take the computer and IT industry as an example, we'll notice that it has been portrayed on film almost from its very inception. The way it's depicted on screen is always up to date, and in many cases *aims to predict* where IT is heading, as exemplified by narratives of artificial intelligence in the recent movie *Ex Machina* (2015) or its predecessors, *AI* (2001) and even *2001: A Space Odyssey* (1968).

Some might argue that Hollywood likes to portray IT in such a compelling and complicated way because Hollywood and Silicon Valley share the same passion for creativity and progress. This is an easy way out. The movies glorify computers and geeks, but they also exalt fast cars and powerful guns.

Another approach would be to look for hidden economic interests shared by Hollywood and the IT, automotive and weapons industries. These ties are not difficult to find and have been discussed in length elsewhere, ranging all the way from conspiracy theories to more sound reports on the rise of content marketing in movies. The most compelling, funny and disturbing report on this marketing trend is Morgan Spurlock's *The Greatest Movie Ever Sold (2011)*.

I'd like to offer another point of view, using other economic activities that blockbuster movies seem to be processing at a much slower pace. Let's start with the world's largest industry, the industry of industries, global finance. Given its unparalleled influence on our lives, the financial industry receives an awfully small fraction of our attention when we go to the movies. The most notable examples are *Wall Street* (1987) in the '80s, and more recently, *The Wolf of Wall Street* (2013) and *The Big Short (2015)*[54], which is based on the 2008 global financial crisis.

In between these powerful images of cynicism, greed and corruption are mostly documentaries, including Michael Moore's *Capitalism: A Love Story* (2007) *The Corporation* (2003) and *The Yes Men (2003)*.

It seems that the golden era of financial documentaries – between the dot-com collapse of 2000 and the financial meltdown of 2008 – has helped us to better understand how *the system* works, but has left us uninspired and, more importantly, clueless about *how to fix it*. It's not surprising that we've come to know Jordan Belfort, Bernie Madoff (real) and Gordon Gekko (fictional) only after the crises occurred,

---

[54] Outside of the silver screen, we can now see the same narrative in the Showtime series *Billions* (2106).

because we wanted to know who took our money and how they did it.

What's truly amazing is how the creative industry's depiction of the financial industry fails to represent (in real time) the technological and cultural changes that finance is undergoing, like the rise of sub-prime mortgages, algo-trading, or indexed funds. We seem to acquire information about what's happening in the financial industry primarily through superficial and fleeting news coverage (*"this is what happened today in the stock market"*), or as portraits of larger-than-life crooks, and there's not much in between.

I think the energy industry is portrayed in a similar way, in that sense that we only get popular culture references about energy during or after energy crises, but never in real time or with predictions for our future. The most popular films show us the newest, most futuristic weapons, the fastest fictional vehicles, and jaw-dropping information technology applications (let's all step into *Q's* office in a *James Bond* film, shall we?).

But when was the last time you saw a groundbreaking wind turbine in a movie? What do we really know about Homer Simpson's work at Springfield's <u>nuclear facility</u>? Even the recent *Star Wars* <u>Death Star's</u> use of an entire sun's energy; is it possible or is it science fiction?

The film *There will be blood* (2007) is a rare example of the film industry attempts of dealing directly with energy issues. The grim plotline is set in 1898, on the early days of the American oil industry, so whatever its creators tried to say about our current energy landscape remains allegoric and ambiguous. Even its title suggests that it is another energy three-Ds story - dirty, dangerous and dishonest. Not an obvious choice for a date night *flick*[55].

The more contemporary *Promised Land* (2012), which describes the impacts of fracking technology on remote communities in the US, provides a great example of the gap between energy technology discoveries and mainstream awareness of their existence. Fracking was first used in the late 19th century and has evolved dramatically since, but its use in large-scale, onshore exploration activities is relatively new, dating to the early 1970s. Fracking is now one of the most important factors in reshaping the world's energy system, but it took Hollywood almost four decades to acknowledge its existence, and to let us know that it's in fact quite dangerous. This lack of pop culture portrayals reaffirms our collective

---

[55] Though it is a great movie with another startling performance of Daniel Day Lewis, which lead to his winning of a second Oscar.

notion of energy – that the industry is dirty, dangerous, and dishonest.

I think the popularity of Elon Musk's enterprises offers the most notable example of the idea that energy is a truly uninspiring technology. While Musk is now almost a household name, ask people to identify Musk's most iconic endeavour and they'll all say Tesla cars. Some might say PayPal, while others will remember his ambitious private space exploration enterprise, *SpaceX*, or his newest *Hyperloop* mass-transportation project. Even fewer would mention *SolarCity,* which he now manages, and has become one of the world's largest providers of solar power equipment, installations and operations. Even Musk's natural charisma and wit couldn't make solar power more popular.

So far, we've never seen the Musk-inspired *Iron Man's* Tony Stark[56] working on the roof with wires and plain black-and-blue solar panels. But we have seen him working on a robotic-war-power suit, which is quickly becoming a close and

---

[56] Though never officially authorized by Musk as his fictional personification, he doesn't seem to mind the comparison, and even made a small appearance in Iron Man 2 (2010)

operational reality[57], and a much more probable technology than a free and abundant energy source.

Hollywood's doom and gloom portrayals of the finance and energy sectors show us that there is something incredibly scary about how we perceive these industries, and this is reaffirmed, not necessarily created, by the entertainment media. We depend heavily on finance and energy for our lives, but we know very little about them – and we only remember they exist when something goes wrong.

It could be said that energy, like all infrastructure, is plain boring, as seen in a <u>brilliant 2015 piece</u>[58] about infrastructure put together by the team of *Last Week Tonight*.

But energy innovation isn't less exciting, nor more incremental than computer programming; when we're doing something we consider "cool" with electric energy, like playing an electric guitar, it can be just as thrilling as driving a car (which is essentially burning fuel), and changes in energy production and its use are incomparably more important to our lives than

---

[57] Williams, How Close Are We To A Real Iron Man Suit? 2013
      http://www.forbes.com/sites/quora/2013/05/03/how-close-are-we-to-a-real-iron-man-suit/#2715e4857a0b710220992f74

[58] https://www.youtube.com/watch?v=Wpzvaqypav8

new weapons (and certainly less sinister). Films like *There Will Be Blood* (2007) and *Promised Land* (2012), but also the surprising, out-of-niche success of *An Inconvenient Truth* (2006) and *Who Killed the Electric Car* (2006) show that there is public interest in learning more about energy issues. These films also disprove any conspiracy theories that energy is kept under wraps because powerful entities want to prevent energy from being publicly discussed. It has been discussed, and there is a wide audience willing to listen.

But these Hollywood examples also show that we keep framing energy as dangerous, dirty and dishonest. Grim and disheartening stories appeal to some parts of our consciousness, but hope and excitement are also very potent and inspire us to act, create and learn.

Mobilizing the die-hard, socially conscious, well-informed, committed activists to raise awareness of climate change was great and led to equally great achievements. But if we want to make energy the stuff of our collective dreams – just like computers, cars and (sadly) weapons – and to inspire us to know and do more, we still have a long way to go.

Part II

How Can We Fix It

or

The Make Energy Cool Blueprint

2. INTRO

From where we stand now, the future of energy is mostly dominated by bad choices. If we stick to our current energy mix, we'll most certainly contribute to a more rapid climate change; increase our air, land and water pollution; and sustain the even bigger, too-big-to-mess-with energy giants.

If we push the paddle on renewables like wind and sun, we'll face some serious technical obstacles, especially in energy storage capacity, and we'll definitely lose precious open space for wind and solar, and possibly many of our wildest and most scenic rivers for hydropower plants. It should be noted that many people, including some environmentalists, don't consider this loss of natural terrain to be a problem, or at least, consider it to be a more manageable symptom of energy generation than climate change.

If we revive nuclear energy, which has zero carbon emissions, we're risking another Three Mile Island, Chernobyl or Fukushima disaster, but on a much larger scale, with tons and tons of radioactive waste, which is worse than any kind of pollution caused by fossil fuels.

Geothermal energy, which is a great source of abundant, clean and cheap energy, can be used in very few locations on earth, including California, Iceland and New Zealand. But in its current state it is far from enough to satisfy the global hunger for power.

Waste to Energy technology (WtE), which converts our refuse into energy, seems like a promising direction; we are definitely generating waste at a rate similar to the rise in energy consumption – and faster than any other generation in history. But so far WtE has yielded a very small amount of usable energy, mainly because it *take*s a lot of energy to *make* this particular kind of energy.

The list goes on, and so far, no technology is risk-free.

We've been through this debate in the past. Not really us, but former generations of regulators, activists and citizens. The energy giants were there the whole time – and we got mediocre results, at best.

Surprisingly, there are some things that most regulators, energy giants and activists actually agree upon when it comes to energy. The first is that most of us can't live without it[59]; the second is that we can consume energy in one place and generate it in another.

The latter convention is so powerful that, until recently, there was a multi-billion-dollar project called Desertec[60], supported by the European Union, with the strategic goal of paving parts of the Sahara Desert with solar panels and providing clean power to Europe. Thus, even ambitious renewables projects accept the idea that energy can be consumed in one place and generated in another (in this case, in another continent).

I think there are two hidden assumptions that we unconsciously agree to when we accept this basic architecture: we accept that energy generation would remain *a system*, and we accept that most of the energy problems and solutions need to be resolved at the system's source. This overall approach could be regarded as *a top-down,* linear approach, which means that the system will feed energy on one end and provide electricity to consumers on the other. Even if the source is different, such as with solar or wind, the same basic

---

[59] Even people who chose to go Off the Grid use some kind of external energy generation, otherwise this Pinterest search would have technically impossible https://www.pinterest.com/search/pins/?q=off%20the%20grid&rs=typed&o=off%20the%20grid%7Ctyped

[60] https://en.wikipedia.org/wiki/Desertec

architecture would remain.

Many argue that instead of constantly trying to increase and improve our energy supply, we should focus on decreasing its demand. This strategy, which has taken many forms, can now be commonly referred to as Energy Efficiency. It is a very important approach and is actually occurring at a much faster pace than supply-side changes.

Greener buildings, energy-rated appliances, improved fuel efficiency in vehicles – all these innovations have been with us for a while, and are happening still with greater vigour.

What's surprising about this *bottom-up* approach, the demand-side innovation, is that we started looking seriously at it only in the last 15-20 years. If we go back to the graph presented in **The Energy Indecision Gap** chapter, we'll notice that investment in energy efficiency is a relatively small and recent development. In the 2000s it didn't deserve its own program and was classified under the generic Renewables and Energy Efficiency investment program[61]. According to the OECD, in 2007, the energy efficiency program accounted for 14% of total American public investment in energy R&D.

---

[61] Though parts of the innovations developed under the Energy Conservation stream, can be regarded as energy efficiency technologies.

What's truly amazing is that while money has been invested in new energy sources for many years and has yielded poor results to date, the money allocated for energy efficiency was spent more recently, but has actually created a remarkable change.

If you visit your favourite hardware store today (real or online), you can buy LED lights, a variety of household insulation materials, programmed thermostats, and many other gadgets and gizmos designed to help you save on your energy bill.

Now go to your favourite home appliances store (real or online). You'll find A-rated appliances in every category, from giant TVs to refrigerators. I bought an expensive, A-rated, energy-efficient fridge about six years ago. Incredibly, the same model would be given a D rating today, due to rapid energy conservation advancements in this product category alone.

In buildings, where nearly 40% of our energy is consumed, there are also new and exciting improvements, such as energy and green building ratings, cutting-edge energy saving technologies, and improved insulation materials.

There are estimates[62] that the global market for green building products (excluding appliances) will grow to $255 billion USD by 2020. This is equivalent to all new investments in renewables in 2012.

The best thing about this growing appetite for green products? Most are totally free-market goods, with few subsidies, government support schemes, or lobbying efforts behind them.

In August 2015, Ikea announced[63] that by September of the same year, all of its light fittings would use LED technology and the company would phase out halogen and 'energy-saving' compact fluorescents (CFL) offered to consumers in all of its stores, *worldwide*. Ikea's announcement didn't occur because the government of Sweden, nor any other government for that matter, was involved in this decision. Ikea, a multinational enterprise with killer marketing instincts and a strong environmental policy, identified a mature and affordable

---

[62] Global Green Building Materials Market is Expected to Reach Around USD 255 Billion in 2020, Growing at 12% CAGR, 2015, http://globenewswire.com/news-release/2015/11/23/789456/0/en/Global-Green-Building-Materials-Market-is-Expected-to-Reach-Around-USD-255-Billion-in-2020-Growing-at-12-CAGR.html

[63] http://www.theguardian.com/environment/2015/aug/10/ikea-ditches-conventional-lightbulbs-for-energy-saving-led-lighting

technology, and used its global purchasing power to make it a more accessible household product and to provide better value for its customers.

When we look toward our *energy future*, it will be interesting to see that at some point in time, in our *energy history*, we actually created a very fruitful branch – energy efficiency. Why is this branch covered with fruits and leaves, while its twin, renewable energy, resembles a small twig?

Simply put: *free markets,* which are fuelled by liberalization, innovation, and inspiration.

Let's see if the free market approach can be replicated on the other side of the supply and demand equation.

2.1 MAKE ENERGY MARKETS COOL

The free market economic philosophy has been with us for a very long time, but its modern interpretation is often attributed to Adam Smith's monumental book, *The Wealth of Nations*[64]. While Smith's opus had many critics (many of whom have never read the original work), it is now the foundation of the world's most powerful economic system. Even authoritarian governments are experimenting with its basic principles. Smith's most celebrated assertion, and the most commonly distorted and debated one, is that by pursuing our *individual interests*, we are in fact promoting the *social interest*, via an *Invisible Hand* that occurs naturally in free markets.

At their best, free markets spark innovation, creativity, and ingenuity. At their worst, markets kill anything that can't be measured monetarily, or require us to be compassionate or forgiving. If you want to dedicate your life to volunteer work, someone needs to fund it. If your newest invention has no revenue model, it isn't worth your time (unless it's an app; that seems to be okay).

----

[64] If you want to read it directly and via a processed version it is freely available here - http://www.ibiblio.org/ml/libri/s/SmithA_WealthNations_p.pdf

I'm providing this short introduction to the free market economy not because I want to convince you that markets are the solution to every problem in our lives[65] or because it is a new economic development, but instead, to suggest that thus far, we have collectively failed to use this very powerful economic mechanism in the energy sector.

Before you shout "privatization" and start that old debate, let me make some key points here. First, there are several places where energy has been largely privatized, most notably in the U.S. and U.K., but that privatization has typically been *vertical* – leading to privately owned power plants, private transmission operators, private plant and grid maintenance services, and more. The liberalization or free markets I'm talking about are *horizontal*, creating an economic system in which *anyone* can generate, trade or sell energy. It's the difference between owning a passenger car and owning an entire rail fleet; the entry level to this latter market isn't universal.

Second, there are very few places in the world where you can

---

65 But there are many who would like you to think so. The most notable scholar on this subject is the Nobel Laureate Prof. Milton Friedman (1912-2006) and other members of the *Chicago School*.

choose your utility provider. But even this rare choice can't be considered a free market, because your options are usually limited to a few providers, as energy is typically a regional monopoly or duopoly.

Third, even if there is more than one provider, there is no real competition on pricing, which is the most important aspect of free market competition. Energy prices are heavily regulated and monitored, which is probably for the best, because it makes energy affordable. But then an energy utility provider can only improve internally by reducing its operating costs. It can't provide a discount to its clients based on these new efficiencies or leverage them to attract new clients. When even significant efforts to become more efficient are not rewarded, why bother?

Fourth, most energy utility companies cannot offer us a personal choice. When I lived in Melbourne, Australia, my utility provider asked me to join the Renewable Service Package, which was a little bit more expensive than the Standard Service. The fine print said that while the company couldn't ensure that my direct energy consumption would come only from renewables, they would invest an equal amount in purchasing and R&D for renewable energy. This is probably as close as it gets to a personal energy choice, but it requires a lot of confidence in the provider. Most people wouldn't buy a car that ran on fuel, while paying for *someone else's* electric vehicle. As consumers, we're used to making our own choices and to enjoying what we pay for by touching, seeing or downright owning our choice.

Fifth, if you know anything about economics, then you'll know that the current energy system is based on Return to Scale, which means that energy produced on a large scale is actually more efficient than energy produced in many different nodes. The larger the furnace, the better its efficiency. This is definitely true. But at the same time, central systems have their own flaws; the large scale actually creates many *inefficiencies*, including base-load generation[66], transmission systems costs[67], transmission power loss[68], and increased security costs[69]. These inefficiencies, along with concerns over monopolistic abuse of energy, are now pushing some energy regulators and economists to design a *less centralized energy system*.

---

[66]https://en.wikipedia.org/wiki/Base_load_power_plant

[67]https://en.wikipedia.org/wiki/Electric_power_transmission

[68]http://data.worldbank.org/indicator/EG.ELC.LOSS.ZS

[69] Some intro into this fast growing cost, associated with central energy generation and distribution can be seen in The Cost of Energy, http://needtoknow.nas.edu/energy/energy-costs/security/

And finally, energy is a state-run monopoly in many countries. In Israel, the National Electric Company (IEC) is a notorious cesspool of corruption, nepotism, inefficiency, and lavish waste. The IEC's current debt totals about $20 billion USD and no one can see the end of it. Because it is such a powerful monopoly with a strong political influence, and energy is such a fundamental necessity, shutting it down is next to impossible, and creating a sizable competition is just as difficult. My energy bill now pays for some very strange organizational decisions, like overbidding on natural gas supply tenders and "forgetting" to update existing contracts to reflect energy market changes. Don't get me started on their parties and wages. The more consolidated an economic activity, the more it's prone to corruption, inefficiency and social unfairness.

So even our most advanced energy markets can't yet provide us with the same benefits commonly offered by other free market sectors, such as innovation, efficiency, choice, competition, or accountability.

The Missing Link

In **The Energy Indecision Gap** chapter, I argued that since the 1970s, energy regulators have chosen energy stability over energy innovation. By default, stability creates stagnation, and in a market environment, stagnation can have devastating results. You need to evolve or you die. Just ask the owners of Tower Records, Blockbuster or Kodak.

This necessary evolution did not occur in energy, because while we can consume content like music, videos and print photos via alternate means, there's no real alternative energy source, unless it's coming from the grid. Actually, there is a small and highly unattractive alternative source that uses complex, unfriendly systems, such as home Photovoltaics, or PV[70]. But in its current form, PV is not for everyone. DIY wind turbines, although relatively cheaper, are also complicated to install, require permitting, and can lose efficiency depending where our homes are located.

As a result of this stagnation and the *non-competitive* market, we are now stuck with an energy sector that is stable, but terminally ill. We have abundant, affordable and accessible energy, but at what cost?

---

[70] Requires initial capital, property ownership and possession of uncommon technical skills (or the ability to pay someone). In many places you still need to pay special fees or taxes to use your own energy, while in other places, installing the PV is actually based on state subsidies.

We cannot resolve our environmental problems with the current energy system, because it depends too heavily on polluting fossil fuels. Our best bet against fossil fuels, nuclear energy, is now *persona non grata* in many places around the world. Thus far, renewables have shown very little progress, and depend heavily on government support and on the goodwill of the fossil fuel energy giants. What's missing here?

I think that a truly free energy market is the missing link. More specifically, it's DIY Energy, or a Retail Energy Generation market segment. If energy generation devices were as easy to purchase and operate as a new car or an iPhone, it would lead to a whole new market segment, a new wave of innovation, and many creative and exciting business models.

The best news? We can start today. In many rural areas, wind is now widely accepted as a smart and legitimate source of energy.

In suburban areas, pending issues such as personal capital, current regulations, and local planning codes, you can install PV panels that can handle most (if not all) of your needs. In urban areas, you can now literally mail order a Natural Gas (NG) generator the size of an outdoor AC unit that would meet the needs of a single household. Unfortunately, they're quite loud, but they can be sound insulated. For a higher budget and for a little more space (about the size of a 40-foot container, or three parking spaces), you can get a natural gas generator that would supply enough power for a multi-unit residential building. And with NG distribution systems already available or planned in many modern cities, these generators won't be too difficult to fuel.

Covering most urban rooftops with PV panels can be mandatory, such as in the case of Israel, Greece and Cyprus's solar hot water systems. Larger buildings can also harness wind-tunnelling effects to generate even more energy.

On a municipal level, we can place NG generators in most public buildings, and also cover their roofs with PV. We can probably use vacant or poorly managed lots for small Waste to Energy facilities. Urban fringe areas can have small and medium wind turbines and even fuel cells. There are many additional considerations, including some serious environmental and economic issues, but the point is we can start today.

There are still two major obstacles preventing small-scale energy production in urban areas. First, it's typically still illegal.

Two separate legal instruments regulate energy production in most countries. The first are permits for production; not everyone can generate the energy to feed the grid and sell to others. The second instrument pertains to the local planning codes, which usually prevent the commercial production of anything in residential or public areas. Yes, we can knit sweaters or make cupcakes on a small scale in our homes, but if we want to distribute them beyond a circle of friends and family, we'll need a license and a space that can *legally* accommodate commercial production.

These interlocking instruments make energy production the privilege of the powerful and wealthy; or of state-run monopolies, with access both to knowledge (to meet or overcome regulatory demands) and property (to place facilities). The only exception to this *regulatory ban* on DIY energy seems to be in remote areas, which usually have a more flexible regime when it comes to renewables (though most places still require permits for equipment installation).

In a way, the current energy production landscape mirrors the American automobile industry at the turn of the 20th century. On one end, there were farmers and miners, rural dwellers who needed automobiles to make a living. On the other, there were the rich and fashionable, who bought cars for style, amusement and fun. And for the rest, the majority of urban dwellers, there were just centrally controlled transportation systems. At this point, saying that cars have disrupted the social and urban fabric[71] is reasonable, but the analogy also leads us to the second major obstacle of decentralized energy, which is public perception.

Anyone who lives in or visits a modern city, from NY to Manila and anywhere in between, knows that cars are a part of the city. But in the early days of automobiles, urban dwellers outright rejected them as being dangerous and dirty, much like how we reject energy generation in urban spaces today[72]. In 1903,W.S. Wilbert, a noble British gentleman, even suggested that London pedestrians should be allowed to freely shoot motorists, for safety reasons.

---

[71] To read more on this subject, see Jane Jacobs' The Death and Life of Great American Cities http://www.amazon.com/Death-Life-Great-American-Cities/dp/067974195X

[72] Most new energy projects are defined as NIMBY projects, or Not In My Back Yard type of projects, with loud opposition, regardless of their proposed technology.

A View in Whitechapel Road, by H. T. Alken:

A cartoon warning against the use of cars in the city, 1831

Source: Wikipedia[73]

It wasn't until the ingenious marketing and lobbying efforts led by early auto industrialists like Henry Ford – designed to make the car a mass and affordable product – that public perceptions began to change. That cultural shift created a legal system that regulated cars in the city, instead of banning them. Needless to say, the proliferation of cars nearly led to a total

[73]

https://en.wikipedia.org/wiki/Effects_of_the_car_on_societies#/media/File:1831-View-Whitechapel-Road-steam-carriage-caricature.jpg collapse of central transportation systems.

The automobile industry bred a commercial and cultural change, which was followed by a social change, which was followed by a regulatory change. Sadly, the energy industry has moved in the opposite direction. At first, energy production was everywhere (and much like cars, controlled by the few), which created a strong social reaction, which in turn led to the creation of a regulatory system that supported energy production only in selective places – an *Energy Ghettos* approach. The same system now prevents energy generation from running more freely elsewhere.

This is not to say that everyone should make their own energy at home *without* supervision or regulation. But just as we've accepted that there are driving licenses, mandatory car insurance, car emission standards, and approved service providers, we can implement a similar system for private energy production.

If we're being honest, cars are still very dangerous and dirty and consume too much of our urban space (about 40%), while creating some serious social issues. But unleashing these machines has sparked great innovation and creative engineering. We now have many different forms of transportation, most of which are essentially mobile *energy generation* units. Instead of converting fossil energy into electric energy, these vehicles convert fuel to kinetic energy, meaning that their engines *move* us, in more than one sense of the word.

We could keep treating cars as if they were just a means of transportation, but collectively, we gave them meaning and symbolism. In the last 30 years, we went through a similar process with water, the simplest of all commodities, when we purchased it in stylish bottles. Water is now a product and a choice – a colorless, odourless and basic human necessity that we have also infused with meaning and symbolism.

Yet we continue to see energy as a plain, dull and unbranded *service*. Thus, we keep consuming energy the same way that it was originally designed to be used, more than 135 years ago, without making it a personal product or a choice, like cars, mobile phones or water.

Use our money elsewhere, give us choice

Many regulators and politicians often talk about creating a *business-friendly atmosphere*. They're usually referring to subsidies, tax cuts and looser corporate regulations. If our governments were serious about resolving our energy and climate issues, they could adopt this same approach, while keeping their precious tax cuts and subsidies (which are actually ours). They could systematically scan national and municipal laws and planning codes to create a favourable, *innovation-friendly atmosphere* for energy. If anyone could generate, buy and sell energy in a truly free market environment, we could come closer to a brave new energy future.

This transition won't be a stroll in the park. There will be powerful interests working to slow and reverse this reformation. There will be a loud opposition with strong (and at times legitimate) economic, social and environmental concerns.

Nevertheless, starting this debate now highlights many of the social obstacles and technical issues that early innovators need to resolve, and that a new legal system would need to regulate. Our current energy system would stay in place, and we would spend very little on structural change, because the shift would be primarily a market-based reform, rather than a governmental effort.

We just need two more components: innovation and inspiration.

2.2 MAKE ENERGY INNOVATION COOL

In the previous chapter we discussed the parallel lines between the energy and automotive industries. There is one more aspect to this comparison; innovation in consumer vehicles is an ongoing effort, with or without government support. Yes, many governments, including those of the U.S., France and Japan, heavily support the automotive industry. But the market laws of supply and demand are working, and companies compete on design, innovation and price. They thrive on choice, not on subsidies. This is why we now have Tesla, and soon we'll have autonomous vehicles. No one ordered this market competition and variety. It just happened, by the *Invisible Hand.*

When American auto manufacturers failed to address the issue of rising oil prices between 2003 and 2009, the Japanese and Korean manufacturers used this oversight to introduce the first commercial hybrid cars (Toyota and Honda), and more fuel-efficient standard cars (Hyundai and Kia). When American makers went big and suburban (Hummer), European companies went to the inner city with hip, smaller models like Mercedes' Smart and BMW's Mini Cooper; the Japanese went with the less stylish but more affordable models, like the Daihatsu Mira.

We can see that when the market is run freely, there's never a one-size-fits-all solution. As consumers, we dislike stagnation. We expect to have a choice and we seek value, and sometimes meaning, for our money.

In contrast, we've become accustomed to consuming our energy just as our parents or grandparents did. Right now, my laptop computer is plugged into a hole in the wall, which draws its power from a municipal grid, which is connected to a regional or national grid, which gets its power from a power plant, typically generated by a natural gas or coal combustion turbine. The great Thomas Edison designed this exact same architecture 135 years ago.

If energy technology were women's fashion, in 2016, women would still be wearing a version of this:[74]

---

[74] We would also probably live in world without whales, which were unfortunate enough to provide their baleen for the period's twisted appreciation of the female body, in the torture devices commonly known as corsets.

Source: Wikipedia[75]

Edison first and brilliantly created a model of this energy distribution system onboard a steamboat, the *SS Columbia,* in

---

[75] https://commons.wikimedia.org/wiki/File: 1898Das_Album6.png#filelinks

1880. He invited his future clients, New York's elite, to see how the incandescent light bulb was superior to the dominant technology of its time, the gas light. This physical Proof of Concept was so powerful that it attracted many wealthy investors and clients to Edison's newly established *Edison Illuminating Company*. Only two years later, on September 4, 1882, Edison switched on his Pearl Street generating station's electrical power distribution system, which provided 110 volts of direct current (DC) power to 59 customers in lower Manhattan.

There are many modern entrepreneurs who wish their newest innovation's Time to Market was just two years, or that their Investors' Roadshow would end over drinks on a cruise ship. Sadly, while we've been smart enough to let women breathe freely and lose the corset, we haven't lived up to Edison's creativity, ingenuity and marketing skills, because we're still using the same basic system design he created. Edison looked at the era's best practice and made it better through a full restructuring. We designed our energy system just like the one on the *SS Columbia*, only on a much larger scale, but we've never taken the next step.

The Quick Green Boat

The government of Denmark, one of the most innovative and environmentally conscious countries on earth, declared in 1998 that by 2050 it would meet 100% of its energy needs through wind power. Denmark has actually over-performed, and on one day in 2015[76], they reached 116% from renewables, which means they could export their renewable energy (or each Dane could take 1.16 hot showers that day), due to powerful gusts on their national wind projects.

Others countries have already produced nearly 100% renewable energy, including Paraguay, Albania and Iceland. These nations enjoy the fortunate combination of small populations and the unique geographic characteristics of abundant hydropower (the two former) or, in the case of Iceland, both hydropower and geothermal potential. Reaching 100% from renewables is a noble ideal, and probably parts of it can be implemented in many other places. Some countries, like Laos and Ethiopia, have a huge, untapped potential for hydropower. Others have enough sun and vacant land for large scale solar projects, like Saudi Arabia and Egypt, and others, like Israel and Singapore, are small nations with limited

---

[76] Read more on this remarkable story here,
http://www.theguardian.com/environment/2015/jul/10/denmark-wind-windfarm-power-exceed-electricity-demand

natural resources and a strong innovation tradition similar to Denmark, so they have both the means and the need to innovate in this field.

At the same time, the 100% renewables approach, like some people so strongly advocate, is a mirage – an illusion of national policy.

If countries were vessels like the SS Columbia, we we'd have four tiny but ultra-sophisticated, slick green race-boats, sailing under the flags of Iceland, Albania, New Zealand and Paraguay. Behind them would be around 190 vessels of various sizes, all traveling at roughly the same speed. And behind those, we'd have two gigantic oil tankers named the U.S. and China. If this fleet tried to make any sudden manoeuvres, like dodging an inevitable torpedo attack (codename: *climate change*) the slick forerunners would get away quickly. Some of the mid-size vessels would make it, but most would probably sink. And the oil tankers wouldn't have even begun to turn before hitting the ocean floor. The powerful whirlpool these two giants would create on their way down would quickly suck in all the other members of this unfortunate armada, including (potentially) the four race-boats.

Just as we can't expect an oil tanker to behave like a race-boat, we can't expect large governments to act like small, fortunate and agile nations. So going to 100% renewables, even if we somehow overcome all the technical difficulties, is a maximalist approach, and probably our best bet for *preventing* any real change in how we look at energy generation.

We have been around this track for a while, and we keep trying to replace the *energy system's sources*. This is a top-down strategy.

Dead Fish Swim Downstream

What if we try to rearrange the system architecture in a new way? Perhaps it's best not to have a system at all. This may sound escapist, but let's set aside the energy resources debate for a minute. We've discussed this issue for a while, but we haven't quite figured out how to improve the system fast enough without losing its greatest benefits, namely stability and affordability. If we can leave the resources issue and focus on the system architecture, we'll realize that there's not much sense in producing energy in one place and consuming it in another. Very few places in the world have the natural geographical advantages required to generate energy, such as geothermal activity or a large river basin, which creates local, clean, cheap, and endless energy.

Fossil fuels are popular partly because they make *energy generation* universal. Australia and Russia may have more coal than most countries, and Qatar may have more natural gas than many larger nations, but these are *tradable commodities*, not *electrical energy,* sold to end-users.

We still need to bring these resources into other regions and build power plants to process them. Coal is heavy, bulky and dry, so it can mostly be used in shore-side power plants after importation or to be conveyed inland at a great cost. In order to export Natural Gas (NG) we need to liquefy it first, by investing a lot of initial energy to cool it down. But in its original form, NG is almost weightless and tends to expand, so it can go underground and inland to be used as raw energy everywhere. From these examples we can see that energy generation is a *question of technology and cost,* not *of geography.*

We can bring energy generation much closer to where it is consumed, instead of sticking to the plan and generating it remotely and in bulk[77]. So if we can make energy generation truly universal, we'll have more resources and more technologies, and far fewer powerful and too-big-to-mess-with energy infrastructure companies and interests. We could have more resources and more options because currently, the energy giants and governments' best interests determine energy resources, which is to have *stable* energy markets.

Once the search for resources was market- and crowd-based, more people and companies would become involved in this activity. Since human ingenuity is an unlimited resource, we cannot predict what millions or perhaps even billions of people could develop. For the first time in our energy history, we could have a *revolutionary innovation ecosystem* of energy solutions.

Buying a $500 kit of PV cells in Home Depot is definitely a start, and so is mail ordering a personal wind turbine from AliExpress or eBay. But we could also have micro NG or Liquefied Petroleum Gas (LPG) turbines, WtE domestic units, personal fuel cells, and many other forms of individual, designed-to-fit, energy-generating units. All of these technologies and others can be made smaller, quieter, cleaner,

------

[77] Some argue that solar is bridging the distance gap, but so far, it has been very slow to perform when we tried to scale it up.

safer, friendlier, more aesthetically pleasing, and last but not least, more affordable.

The first car designed for the masses, the Model T, was spartan and functional, or as Henry Ford famously quipped, "any customer can have a car painted any colour that he wants so long as it is black."

The first real consumer mobile electronic device, the transistor radio, was clunky and had limited functions (only one band). After several waves of innovation, fuelled by economic demand and consumer excitement, we now enjoy many more variations and options in our driveways and pockets. The Tesla and the iPad owe their slick designs and improved interfaces to their early, clunky and unattractive prototypes.

Let's start making the PV, the micro-wind turbine, and even the huge, fossil-fuelled power plant nostalgic but irrelevant prototypes. If there were a real market for their predecessors, these too could look like vintage artefacts.

The Energy Open-Source Community

There are two more aspects of energy innovation that are almost impossible to avoid in a book written in 2015/16. The first is the buzzworthy phrase, *Digital Disruption*. The most notable examples of this disruption are Airbnb in the hospitality industry and Uber drivers competing with taxis. Essentially, digital disruption uses digital information platforms to create competition in places where competition went dormant, or at least where it followed a single business model.

To provide personal, on-demand transportation, you used to need a taxi, a dispatch radio, and a license. Now you just need a car and a mobile phone. In the hotel business, you needed to have a brand, a booking system, a concierge, and an expensive, multi-room building in a prime location. Now you just need an extra bed and an app.

Like so many buzzwords, the hype and excitement around digital disruption prevents us from seeing where and how it works, and more importantly, where it doesn't work. Most importantly, digital disruption takes a basic need, strips it to the core, analyses how else it can be provided, and then creates or leverages a digital platform to deliver the same service. Sadly (or luckily, depending on your personal interests), the energy sector is currently immune to this type of disruption, because the means of providing an alternate service is still unavailable.

Think about airlines. There could be many different ways to book a flight, but in the end, we still need a pilot, a runway and a plane to provide the service. Only if someone who already owns a plane would offer us a direct service, bypassing the standard airlines, could we enjoy a real disruption. Unfortunately, the people who currently own planes don't like sharing their ride.

The *Energy Airbnb* can't emerge yet because, as we've seen, unless you own a power plant, you can't offer a competitive service, nor a better price. Only when we have a decentralized energy system, a physical disruption, are we likely to see a digital disruption[78].

The second type of modern innovation, and the one I find the most relevant to energy, is the Maker, open-source or DIY revolution. R&D used to be an organizational or an institutional effort, with big costs and years of meticulous work, sometimes leading to great success and sometimes to costly dead-ends. Today, the world's largest R&D labs live in

---

[78] This doesn't imply that we can't disrupt some parts of the current energy system. For example, the way energy is currently billed is totally irrelevant to the 21st century. We consume energy with virtual units (kWh), without knowing its real costs and we pay long after we use it, regardless of the value it provided. With the exception of municipal water, there's no other product we consume and pay for this way. It is the same system that Edison has designed and it no longer makes any sense. If digital disruption could start anywhere in the energy sector, it should target the billing systems (which sadly have nothing to do with energy generation, energy efficiency or climate change).

people's basements, garages, attics, and personal computers. Yet, the garage, bootstrap or lean startup model has actually been with us for while. The Wright Brothers invented aviation this way, even though Samuel Pierpont Langley, a well-known engineer and inventor of his time, tried doing the same thing at the same time with much deeper resources and the blessing of the American Ministry of Defense. The Wrights changed the course of history because they went lean, smart and ambitious, while Langley's investors played it safe and lost a fortune in the process.

But personal R&D is now a much wider and more influential phenomenon than we seem to understand. As I write this, there are millions of people dreaming of making it big, and they actually can do something that might change the world. They have much better means than the humble Wrights. They have 3-D printers, affordable power tools and parts, and coding knowledge, but most importantly, they have access to the information superhighway – and therefore to each other.

These makers don't need anyone's permission to do their work. They don't need someone to approve extra budgets or additional staff. They report to no one but themselves, and the market, when and if they think they're ready. There's no protocol for what they're doing and there are no rights or wrongs. Check out James Bruton, from Southampton, U.K.

He's not selling anything. He just loves robots, 3-D printing and coding. Bruton is making real-life robots from scratch, for fun, spending hundreds of hours on his hobby, and in the process, inventing stuff that would make many R&D professionals' heads spin[79].

I'm quite sure that some of these makers are working right now on the next energy revolution, with a tight budget, but with great vigour, talent and enthusiasm. Their dream probably includes selling their invention to someone, instead of starting their own energy utility company, like Edison's. That's a noble effort, and I'd be the last person to discourage innovation, but if you are one of these amazing people, please make sure that your business model *isn't* based on selling your creation to an energy giant or a government official. They'll do everything in their power to bury or own it, just to bury it later. And it's not because they're bad, cynical people, or because they don't want you to succeed or because they don't care about climate change; in fact, they might be the exact opposite. But they have been trained, programmed and ordered to keep the energy system in its current position – safe, stable, and centralized.

So carve your own path, make your own business model.

---

79   See his site here, http://xrobots.co.uk/author/

Enable us to see your newest energy generation innovation just as we would see a new car or consumer electronics innovation. Make it personal, aesthetically pleasing, user-friendly, and interactive.

Those of us who care enough but can't make anything useful can do something else to help those who can. We can inspire them.

2.3 MAKE ENERGY INSPIRATION COOL

Humans can survive 21 days without food, three days without water and almost eight minutes without oxygen, but we can't live even one second without hope.

Proverb

Of all the ideas in this book, I believe this chapter is the most important, because without inspiration, there's not much we care to do.

As noble as that proverb sounds, I think we tend to give up on hope too fast, but we're not really saying it out loud. We just go on with our lives knowing that *some things will never change.* But as a wise man once told me, "*never* is a word only young people use."

As we grow up, we learn that taboos soon become the norm, and solid, age-old structures can crumble like castles made of sand. Being openly gay was, until very recently, illegal in many countries. Sadly, in some places it's still illegal (or still a taboo), but we are now surrounded by openly gay people and enjoy a thriving queer culture. We also order products from unknown vendors on the other side of the world, which would have seemed a risky business only 20 years ago. Now, my 72-

year-old father does it several times a day. Publishing a book, like this one, used to require years of research, prestige and the consent of a publisher.

We like to think that technology has made all of these changes a reality, but we tend to forget how inspiration contributed to the process. Being gay is not a choice, but being a free person and making your own choices is a basic human right, and this notion inspired many gay people and their heterosexual, LGBT advocates to battle discrimination.

By contrast, globalization and the free flow of ideas was a long-time dream shared by many, but it took several millennia for humans to stop seeing what separates them and to start considering their mutual interests. Trade and creativity are two of those connection points.

The actual source of human inspiration was for many years a mystery. The ancient Greeks believed in the Muses, or mythical women and beings that inspired artists and philosophers to be more creative. Several other cultures across India, Asia and Africa share this idea, which was likely inspired by a single, unknown, primordial source. Judeo-Christian culture has attributed the powers of creativity to the presence of divinity in all of us. And since the 16th century, people have been discovering that the riskiest, but sometimes the most exciting route to the gods and deities of inspiration is lined with a dizzying array of mind-altering substances.

But in the 20th century, humanity started working on a fantastic new system for creating inspiration. This system, which I will call the Inspiration Machine, is so ingenious that it doesn't have a single form. It comes in many shapes and sizes, including but not limited to comics, games, books, music, TV, film and art. It also draws it powers from various sources, and literally anyone can contribute – a blind albino singer from Mali like Salif Keita, an American drug addict like Jim Morrison, a British single mom like J.K. Rowling, or a South Korean rapper named Psy. The Inspiration Machine's greatest achievement is its ability to spread its creativity everywhere and draw even more people to participate in its quest for extraordinary talent and the next inspiring project.

In **The Energy Inspiration Gap** chapter, I showed that pop culture typically frames energy matters in a rather grim and uninspiring light. In **The Energy Information Gap,** I also explored why this framing is matter of inertia, rather than a strategic or conscious choice. So writing, drawing, filming or developing a new creative project that would *inspire us to see energy positively* is difficult, because it rarely happens. But you would expect creative minds to find a way to do this just because it is difficult. Or rather, to "borrow" a way.

In the words of Pablo Picasso: "Good artists copy. Great artists steal."

Let's Go Stealing

Creating a ground-breaking creative piece that changes how we perceive energy doesn't need to start from scratch. Similar projects have reshaped the way we think about other technologies and human efforts. I'd like to suggest three, similarly inconceivable creative projects or campaigns that have succeeded beyond the original creators' wildest dreams.

The first example is, of course, cars. Originally designed to replace more humble means of transportation, cars became the 20th century's dream machines, thanks to what is probably the best marketing campaign in history. Cars emerged as a popular artefact side by side with the emerging motion pictures industry, and owe at least some of their success to this visual medium. When we think today about cars, (depending on our age), our archetypal images often reflect the cars driven by James Dean, James Bond, Steve McQueen, Thelma and Louise, or more recently, Vin Diesel. These toxin-emitting, life-threatening, space-consuming and congestion-creating machines have captured our hearts and souls in such an effective way that we can hardly imagine living without the freedom they provide us (that is, the freedom to keep buying them and paying for their precious fuel, insurance, repairs, and parts). Even people who don't own a car know something about tires, fuel consumption, batteries, steering wheels, and wiper blades. This is more than most of us know about energy generation systems.

And our experts? Wow, the stuff they know: torque, horsepower, acceleration. You don't even need a job building or designing cars, you just need to love them. Notably, we don't get our information about cars from watching the news anymore. Just close your eyes and imagine the three hosts of *Top Gear*, the world's most popular TV show, surrounded by hundreds of people in the audience, typically men, all taking ecstatically, passionately, and personally about each car's RPM. This scene, which we now take for granted, means that vehicles have left the realm of pure technology or industrial production and now provide many people with a passion that transcends the need for transportation. Cars are something that many of us long for and dream about.

So here we go:

Hypothesis #1 - when we love something, technicalities don't really matter.

Hypothesis #2 - when we love something, we don't care if it's loud, dangerous or expensive.

Food preparation is the second example of popularizing a technology or activity. While we live in a time where TV cooking shows parallel the Bible in Middle Ages society (in the sense that it seems to be the most important knowledge to acquire), it's interesting to remember that we weren't always tuned into food TV. Less than two decades ago, food preparation had two extremes: *nouvelle cuisine*, in which the best food was prepared by the world's elite and eaten by the wealthiest people, and simple, Mom & Pop-style diners. Recipes were passed along to housewives through daytime TV, weekend papers, and less-than-inspiring cookbooks. Unless you had three Michelin stars under your chef's hat, you were just a cook, not an undiscovered culinary genius – and it was a pretty dull and uninspiring task[80]. Most people used to go out for meals when they ran out of food (low end), or to celebrate a big occasion (high end), but the food came second, after the sense of social occasion. Few people shared the idea that eating was a cultural experience, like that of movies or literature. There were very few authentic "foodies."

The late 1990s and early 2000s brought a tide of more modern, yet accessible cooking enthusiasts, who I think are

---

[80]   If the film *Chef* (2014) were set in 1984, Jon Favreau's character would be a middle-aged man without money, a steady job, a loving partner, appreciation from his kid, or a future of any kind. He could never get the girl, the crowds, or the money. It would have probably been a critically acclaimed, art house tragedy.

personified by Jamie Oliver. The Naked Chef made food preparation beautiful, fun, engaging and, let's admit it – cool. It helped that he was not very wealthy, and he was genuine, good-looking, and had a charmingly subtle lisp, which made him slightly more lovable than many of the ego-maniacs who seem to have dominated this scene before his arrival (and sadly, ever since).

Most notably, the act of preparing food, even at home, has remained pretty much the same. It is still time-consuming, physically demanding and messy. As a profession, cooking is still underpaid, unstable and extremely abusive. Only its image has changed. Preparing food is now an art form, with low and high ends and a lot of shades in between, just like any other medium.

Hypothesis #3 - when we love something we will keep doing it – even if it's difficult.

Let's consider coding. If you're a coder or you know one personally, then you know how dull, tedious and irritating it can be to do coding work. Even today, though your dear old *nana* probably has an idea for a great app, it seems that the number of programmers who are looking for a way out exceeds the number of people looking for the way in (it's mostly the high salaries that keep them in). This is not to say that coding isn't important, or that it can't be fun (it can be amazing); it's just that the image of software development is slightly more glamorous than the everyday tasks it involves. So what made coding so appealing? Four words: Bill Gates, Steve Jobs. But if we're going to choose just one word, it has to be Google.

Hypothesis #4 - when there is a great personal and social potential, boring things become cool.

Politically Incorrect

I'd like to add a final, politically incorrect assumption. Most successful, creative movements and their marketing campaigns have all targeted a specific demographic group: young men. Cars, cooking and coding's greatest role models always seem to be young men. Typically, their earliest manifestations were young Brits or Americans. This shallow observation, of course, says nothing about people from other countries or women's contribution to the success of these changes. It also says nothing about the power of Anglo-minorities to participate in the effort. Steve Jobs came from a Syrian background, Mark Zuckerberg is Jewish, and Google's Sergey Brin is a Russian-American.

But if we examine these examples, we'll see that the specific poster boys (and their backgrounds) don't really matter. Instead, their legacy is what endures. Cars have made it to the North Pole (again, thanks the *Top Gear* crew) and everywhere else on the planet, food preparation is now a global obsession, and coding, well, it made us all connected.

Hypothesis #5 - it takes an English-speaking young male to make something look cool.

Before you throw away this book in a rage and accuse me of trying to perpetuate a white, masculine and Anglo-centric stereotype, let's see how this formula worked thus far – but we'll soon give it a genderless spin. Following Joseph Campbell's *Monomyth Theory*, the popularization of technological developments usually follows a distinctive pattern.

At first, technological innovations are the brainchild of mavericks, like Henry Ford and Thomas Edison. Then they slip into popular culture with the seemingly unintentional help of a young, often humble but free-spirited male (James Dean); this protagonist's use of the technology seems effortless, but at the same time subversive (Steve Jobs). Other young men, who are typically early adopters of new technologies, see this cool, subversive attitude and want to be like him. Then the rest of us discover the innovation and want to do the same. The chances of a successful narrative are greater if this person is fluent in English.

I mentioned Elon Musk's rise to stardom in **The Energy Inspiration Gap** chapter. It seems that his rise, much like Virgin's Richard Branson a few years earlier, has inspired the creation of a new type of hero – *the sole tech entrepreneur*, *fighting for his principles and a better world through great force and ingenuity, against villains and apathetic bureaucrats.*

This, too, isn't really a new genre, though getting rich as an entrepreneur seems like an upgraded version of Bruce Wayne, who inherited a fortune that enabled him to become the **Batman**[81].

So if our protagonist is committed to a single idea or better yet, to his own enterprise, like Jobs, Oliver or Musk, and he is fighting the odds, chances are that more young men will identify with, envy, and try to duplicate his success. Then the rest of us will follow, too.

Real Girl Power

So we now know roughly how a new technology filters through our popular culture. But since this book is about the future, or more specifically about a future created by many people rather than by a few privileged individuals, we can explore how this future could represent more current and inclusive social norms.

We now have more examples of fearless female heroines, who

---

[81] This new type of hero is an interesting comparison to what many people consider to be the greatest movie of all times, Orson Welles's Citizen Kane. In this film, inspired by William Randolph Hearst's biography, the protagonist is becoming rich beyond his wildest dreams, though Welles's portrayal shows a man who lost his moral compass, not a man fighting for his principles.

represent a different type of protagonist: Khaleesi in *Game of Thrones*, Katniss Everdeen in *Hunger Games*, or Amberle Elessedil in *The Shannara Chronicles*. *Star Wars: Episode VII - The Force Awakens* (2015) introduced us to the newest Jedi Knight, a confident young woman named Rey. All these figures represent a new breed of hero, who is not only feminine, but depends less on physical strength (though you wouldn't want to pick a fight with any of them) and is much more open and collaborative. Khaleesi consults with her male advisors and guards, but controls her own destiny. Rey cooperates with her friend Finn and the aging Han Solo, but she seems more determined than both of these men.

If we combine this new protagonist model (the benevolent, courageous female leader) and the purpose of this book (new technology for a world that desperately needs it) we can create an interesting, technological twist on the *Monomyth* as we've come to know it.

All the female examples we've seen here operate in a single genre, which is fantasy or dystopian sci-fi. While powerful female role models are now more common and, arguably, more socially acceptable, we still haven't found a way to portray these figures in a more relevant and contemporary context. We want women to have better opportunities, to be more powerful and hopeful, but in order to actually visualize these changes, our filmmakers and authors need to create *an entirely new world.*

So let's bring it down a notch, by letting women shape the very near future or even the present. Let's encourage women to invent, design and create real and necessary change, whose time has come. This could be a win-win situation – women leading a positive change through new technology; and an emerging technology accelerates a positive social change – not in a galaxy far, far away, in world of dragons, nor in some hostile dystopian environment, but in our world and our lifetime.

Needless to say, our energy protagonist could be a young child, an elderly person, an Indian or African entrepreneur, or any other socially or technologically excluded group. I'll be delighted to see a group of people from different backgrounds working together to change our energy future. If you can create a more effective narrative with any of these options, go ahead and do it. Make it colorless, genderless and ageless. Make energy generation look effortless, fun, and inspiring.

Just make it look really, really cool.

PART III

EPILOGUE: BEST / WORST CASE SCENARIOS

3 INTRO

In this final part, I want to be my own devil's advocate. There's something terribly tempting and self-indulgent about presenting your own ideas and finding them to be true.

So, I'm going to challenge the major ideas presented in this book and see how they stand in case some or all of my basic assumptions cease to exist. I'll try to be as honest as I can, though I'm sure that there are many more flaws in this theory.

But before we start, let's begin with some optimism. Let's see how our lives could look once the ban on retail energy generation has been lifted and the new market emerges.

3.1 Best Case Scenario: The Internet of Energy

The best news is that our lives, our homes and our communities would look exactly the same. We would still wake up every morning, drink our first coffee or tea, and go about our daily activities at work, at school, or in our free time. Our homes would still have all the HVAC, lighting, electrical appliances, and hot water we've come to appreciate. Most of us won't work in energy production, and energy will still work for us.

The first real change would be in the electrical appliances themselves, and the hardware stores (real or virtual) we already know. We'll need a new aisle to display the state-of-the-art energy units. They won't look like those clunky, noisy and polluting diesel or NG generators that we have now, either in construction sites or as standby units. They'll be enclosed in compact boxes, about the size of a small AC outdoor unit. There will be various prices and designs, but we could assume that they'll cost between $500 and $3,000 USD, pending the design, capacity and underlying technology. There will be a few leading brands, provided by early-movers and fast-thinking technology companies, but there also be hundreds of designs and manufacturers from all over the world, including developed nations, because it would be a high-tech unit, not a 'dumb' generator.

After we chose our unit (let's call this one *The Force*), the store would arrange for its delivery. A qualified technician would install it in our house. It could be mounted on the roof, in the yard, or next to the washing machine. We wouldn't really care, because it'd be silent and produce zero or near-zero emissions.

The Force would feature a simple NG inlet[82] and an electricity outlet. If your house were already installed with PV, a micro wind turbine or a WtE device, you would connect those to *The Force*, which would serve as your household energy manager.

Operational and mechanical features are the least important part of *The Force*, which would embody the Internet of Things. In fact, it would be the *Internet of Energy*. The basic ideas

---

[82]   Sadly, at this point my scenario hits the first blow of reality. We need a
technology that is universally available, doesn't require special skills and
doesn't require permits. Of all the available resources, natural gas is the most
common, the safest, the lowest-carbon and the most affordable choice. There
are definitely cleaner and quieter options, but they don't quite fit these three
requirements. Fuel cells are extremely expensive, WtE output is still not
enough, wind (and in many place, solar) may require permits, and they
require you to own the lot and/or roof, so they are not an option in the dense,
built environments where most people live.

behind the so-called *Smart-Grid*[83] would occur on a much smaller and finer scale. Your household energy providers – the municipal grid, the PV, wind, WtE and our new *Force* unit – would all be managed from a single processor, using a sophisticated algorithm designed to optimize our energy consumption. The municipal grid would send continuous tariff data, much like smart toll roads calculate fares based on supply and demand. When most of us are at work, the grid electricity for domestic consumption would be at its cheapest. When most of us are at home, it would reach its highest levels. Our other units would also send their signals. When it's windy, our turbine would indicate that free electricity is available, and when the sun is out, our PV would indicate that it, too, could be used. When our PV and wind batteries start to lose power, the system would decide whether it's most economically efficient to charge them with the grid or *The Force's* own power. If our neighbour bought the expensive, larger capacity and more efficient *Force III*, and we set *The Force* unit to communicate with it, *The Force III* could also offer us cheaper energy.

But none of this would be our concern, because *The Force*

---

[83]   Smart grids are an improvement over the existing municipal grid because they can connect power both downstream (e.g. feed power to consumers) or upstream (e.g. get end users' electric output). They are much like the Internet, because they allow input (data/power) and send output (data/power). Most modern grids are designed to work downstream only.

would manage these choices internally and automatically, unless there was a mechanical or electronic problem. In this case, *The Force* would let us (and its service provider) know about the issue through SMS or app notifications. Moreover, we'd still have the municipal grid as backup.

But the best feature of *The Force's* algorithm is that it would manage our demand-side as well. With the Internet of Things, all or most of our energy-intensive appliances can be monitored. If we press the "start" button on our washing machine, dryer or dishwasher, the system would analyse and determine the best time to begin its operation, which might be two hours after we leave the house, when energy is expected to be cheaper. We could always set *The Force* to operate in Manual Mode via the app, or override this automatic function on its touch screen.

The Force's algorithm would also learn and analyse our energy-use patterns. When we get close to home after work or school, *The Force* would prepare all our energy needs. Our HVAC systems would be turned on to create our favourite atmosphere, but optimised to invest as little energy as possible, rather than the quick cold or hot thermostat peaks we usually experience on arrival. If we like to take a long, hot shower at the end of the day, this, too, would be optimised and synched with our hot water system. Light sensors would also be integrated into the system.

There would also be books about energy units, plus user magazines and websites, and a vibrant second-hand market. Both you and your neighbour would eagerly await the launch of the mysterious new *Force X*, which will inevitably be the biggest thing to happen to energy and free us from the NG grid, too.

But the real changes would happen on a much larger scale.

The power of choice, the choice of power

The large energy stations – the energy ghettos we've come to know, fear and loathe – would change their appearance, technologies and business models at a rapid pace. Their furnaces would be divided into smaller, cleaner and safer technologies, and they would look for opportunities to move closer to the demand. Because more of us will know first-hand that living next to your energy unit isn't so bad, we'll be more tolerant about new energy projects. Just as living next to major, inner-city public transportation routes was once a real estate disaster, those apartments are now the most desirable.

The huge, old *power plants* would become obsolete and the new *power houses* would barely resemble their primitive, gargantuan and dirty ancestors. They'd be slick, with a state-of-the-art design, and feel quiet, safe and clean. They'd have to be safe and clean, because they'd be much closer to where people live and work. There wouldn't be any trucks coming in or people arriving at their gates for 24/7 shifts, because most power houses would be 100% automatic, and centrally controlled via sophisticated municipal control rooms – like the ones we have today for water or transportation infrastructure. Once a day, one or two technicians would arrive with safety vests, helmets and tablets to check their systems and safety. There would be several equipment providers, municipal operators and technicians working on these units, rather than vertically integrated regional monopolies.

There would also be specialized utility providers. Some would target retail consumers, while others would target municipal and public consumers, and they'd all compete on small- and medium-size factories. The municipal grid would be a smart-grid, because there would be millions of users who are constantly seeking an optimised price.

This is where the energy companies can really go high-tech. The principles of Return to Scale are mostly economic, but they are also thermodynamic. If energy firms have the biggest equipment and turbines, and the biggest resources purchasing power, then their output is bound to have the *lowest price*. If they make their HR, control and distribution systems as efficient as possible, they can provide us with cheap energy from their rural, suburban and inner city facilities – based on their own algorithms. If they also owned large-scale renewables facilities, they could factor those in too, or provide us with a renewables-only Premium Service.

At this point, you might think that the large utility providers would use their superior scale, resources and price to crush the little guy, or the system would experience consolidation and we would go back to square one. It is a possibility, but I'd like to explain why I think it's implausible.

While the price signal is important, in the sense that we typically prefer the lowest price, reducing human behaviour to nickels and dimes is simplistic and empirically inaccurate. Price isn't the only thing that matters in our personal choices. We are emotional and social creatures and we react to symbolic and social signals, not only to prices. Apple's iPhones are the price leaders in every country worldwide, but they're also the highest-selling smartphones. Apple's ingenious marketing strategy, embodied in its unforgettable tagline, *Think Different*, has made its products both mainstream and subversive, upmarket but universally desired. Apple, first through their computers and then through their mobile devices, has allowed many people to identify with the idea that *they are different*. Apple then provided its clients with the *freedom of choice*, and the freedom to buy the most expensive product in the shop (after standing in line for hours with *similarly* different people).

I think the same social and emotional levers could operate with energy consumption. Once some or most of us realized that energy could be *owned* and not just *paid for*, we would want to keep this option, much like we prefer to own our own car (even though it's clearly uneconomical). And unlike in transportation, we wouldn't leave our service providers because we would still need them as backup. The rail and bus companies haven't been dissolved[84] as a result of the automobile, they just provided a different kind of service to a different market segment on a different scale.

The new system architecture would create two major improvements over the current one. First, it would bring energy generation closer to optimum, because there would be a far higher correlation between supply and demand. Second, it would create a much stronger economic incentive to reduce emissions, because both private and commercial units would operate closer to where people live, study and work, and we would only agree to have minimal emissions and noise levels in these places.

I want to mention one final, but crucial point: all of this won't cost a dollar for taxpayers. With the exception of some legislative efforts to make widespread energy production and ownership legal[85], we don't need to push subsidies, public

---

[84] Sadly many monorail, trams and streetcars services have been decommissioned. There are various theories on what lead to their demise, but there's some evidence that at least in several U.S. cities, this has been the result of deliberate action by the auto and parts manufacturers to minimize competition.

[85] We pay for our energy regulators' time anyway, but currently most of their time is spent on *preserving* the current system architecture.

R&D or public campaigns. These are all market goods and services.

The only two aspects that will require some central intervention, but in a later stage, would be to ensure that energy remains affordable for those who cannot own either property or personal means of production. There are several ways of structuring this assistance, but we're not there yet. The second government, or rather municipal, involvement would be to make some small but necessary adjustments in the local grid, in order to increase its intelligence, to a Smart Grid. These changes and their costs are likely to be minor in comparison to rebuilding our entire energy system, or replacing all of their resources. Both of these small interventions would come from taxing the huge economic activity created by the new DIY energy market, and they could kick in only after we see some activity happening. It could actually be a financial *net benefit* result to national treasuries, instead of a cost.

Before we continue to the Worst Case Scenario, and before you pull out your pen, paper or keyboard and crush my sci-fi, flowery Best Case Scenario based on all the economic, environmental, thermodynamic, mechanical and legal impurities and flaws you've found up to this point, I would like to suggest a better use of your time.

Take the same pen, paper or keyboard and start designing your own energy system. Map out where and how energy is generated, who owns it, how it's sold and where it's consumed. Ignore reality, the laws of physics, the laws of men, my ideas, and if you can, your own best case (100% renewable / all-you-can-emit-carbon party / and everything in between). Start from scratch.

Try to imagine an optimal system, including its components and basic architecture. Make a couple versions. Don't think of how *it could* look; think about why and how *it should* be designed. There's a huge difference there. Create your own sci-fi, visionary, or whimsical system.

Launch your own startup company based on these ideas, or share it with the rest of us in www.makecoolenergy.com or on Facebook. Let us know what you think we can do to improve our energy future, not what cannot be done.

3.2 WORST CASE SCENARIOS

In this section, I would like to explore what can go wrong. It's important to realize that unlike what we typically see on TV, in movies or in books, there are no guarantees or happy endings when a large-scale technological change or disruption is occurring.

Even if the COP 21 commitments could be implemented right down to the punctuation, we'd pay an unknown price for these changes. This price is unknown because large systems, like national and global infrastructure systems, always operate at an equilibrium point – an optimal operational mode where all the risks and benefits are known. Many people and organizations economically, socially and often legally support this equilibrium point and, when it's shifting, we can begin to see the actual winners and losers.

Many years ago, not long after dinosaurs roamed the earth, phone calls between people were possible only through landlines. People knew they could have one phone at home, one at the office, and occasionally use one on a street corner. But when cell phones started to spread, we suddenly had other options. We could have a cell phone, and still keep the home and office lines, or we could cut the cords and never again rely on that corner payphone. Traditional telecom companies, like Bell and British Telecom, started to see some revenue drops, which placed them on the losing end. Others, like mobile operators Verizon and Vodafone, and equipment manufacturers like Samsung and Apple, came out on the winning end. This change has created ripple effects, too. A small, Danish/Swedish/Estonian technology company called Skype realized that long distance calls shouldn't be exclusive to land lines, but instead, use Voice over IP (VOIP) technology to make calls. The telecom companies took another blow.

These telecom giants had to work hard to provide their clientele with better offers and push internally to become more efficient. They didn't collapse, because we're still using landlines, and they have remained profitable. But their overall importance has declined.

The fear of a technological change is well known and understood, but the fear of stagnation is just as powerful, so change can be managed. Let's dig a little deeper.

There are many assumptions behind this book's core theory that energy generation and trade could become more universal and popular, so I've divided this meta-scenario into a smaller set of scenarios. I've listed them from the most to the least probable scenario, though we may have our differences here, too.

Worst Case Scenario #1: No buy-in. People still hate DIY energy.

Before we continue with this scenario I'd like to offer another view on this matter, which hit me on the head during a recent trip to Germany. In most modern economies, about 45% of domestic energy is consumed for heating and cooling spaces and ventilation (HVAC) and another 18% for hot water, which completes the lion share of domestic energy needs[86].

People who live in a warm climate, like me, typically use electric energy to cool and ventilate buildings, and in some cases[87], to provide hot water. But people who live in cold climates, like the first world residents of Northern Europe and colder parts of North America, consume energy much differently, via their heating systems, which are usually NG based (which in many cases is also used to provide hot water). While this seems like stating the obvious, it's actually a primitive, lighter version of what I'm suggesting here. These systems are actually energy generation units, but they convert fossil energy (NG) into thermal energy (heat) and these

---

[86] To read more see the American EIA Residential Energy Consumption Survey (RECS) here http://www.eia.gov/consumption/residential/

[87] Several warm climate countries like Israel, Cyprus and Greece have mandated the use of solar hot water system after the 1970s oil crisis, so people living in these countries usually don't consume electricity for this purpose (as long as the sun is out and for a short time after).

systems are typically 'dumb' as they burn fuel at a constant rate, and are regulated by thermostats or manual settings. This is basically what occurs in a power plant, which converts this heat, albeit on a much larger scale, to electric energy. This last form of energy is fed into the grid and then redistributed to our homes, workplaces and schools. So, many people are *currently* making energy in their homes, but they don't think about it this way. And they don't seem to mind, or rather, they see this kind of energy generation as an essential part of their lives.

But let's assume that generating heat and generating energy are fundamentally different and require us to step way out of our *comfort zone* or in this case, out of our *thermal comfort*.

Failing to achieve energy buy-in is a most probable scenario, because for a very long time, we've been taught to avoid seeing how our energy is produced (much like we don't want to see how our favourite sausages are made).

In this scenario, we've added legal conditions that allow people to own their energy generation units and to sell their output privately. Some die-hard fans, energy geeks, if you will, have adopted this new development with suspicious and overly joyous enthusiasm to create a nice, stable market niche, like organic beer. They have their own web forums, specialty shops, strange t-shirts, and the occasional trade show. You won't be able to buy slick, cutting-edge energy units in Home Depot, but you could get some imported supplies and rambling advice from a weird-looking guy with a beard, who left a handwritten note in the craft brew pub. You'll become Facebook friends for life.

As we speak, there are reports of people doing exactly this – living in the inner city but going off the grid long before it's socially or legally accepted. They could be avant-garde radicals, leading us into the future. Or they could be a modern version of the Amish, who have been living *off-grid* for 200 years. But this fringe lifestyle choice also challenges us: do we really want to have another eccentric group who views energy as a satanic technology, or do we want elements of this independence to go mainstream?

The Silver Lining: we can all go back and continue our debate on fossils versus renewable. We haven't spent a dime on these weirdoes.

Worst Case Scenario #2:

Innovation doesn't happen. We have the same energy options.

I consider this scenario to be slightly improbable, because once we have millions of consumers with different needs and circumstances, we won't be able to use only 3.5 resources (wind, solar, NG and the laggard WtE) so more options would eventually turn up. But even if this assumption is wrong, then wind and solar are bound to be more popular and easier to use, because they'll be the cleanest options and they would have to compete against the easy, evergreen and reliable new units. Less stringent regulation would make all technologies easier to own and more abundant, and would eventually would push prices down and increase innovation – even if only within each individual technology.

The Silver Lining: wind and PV would go out into the world without subsidies, complex setups and permits. Prices would go down and reflect true market expectations and demand. Some fossil activity would move from the energy ghettos to the inner city and suburbs. It would make the system more resilient and reduce pressure on municipal grids. Utility providers would focus on medium and large consumers, where demand is more predictable and manageable.

Worst Case Scenario #3:

Emissions won't go down.

There's a chance that even if we implement the steps I've suggested on a large scale, it won't change the landscape quickly or drastically enough. We will have shifted from large turbine emissions to smaller, scattered turbines. This case implies that we'd have more emissions in densely populated areas, which is something that we're currently working to minimize. However, I find this to be a very improbable scenario. I'm skeptical about this one, not because I don't like its consequences or that I find it difficult to untangle, but because I don't see us knowingly buying something that would make our lives *worse*. My theory would only work if we were offered a no-brainer unit – one where we didn't think about its operations, noise or emissions. We'd live right next to this equipment, so it would have to be free of major irritants.

Whether the unit would have some kind of internal carbon capture, a fuel cell, or any other technology that reduces emissions and noise to a bare minimum, we wouldn't buy it unless it was safe and made our lives better.

Once the market has been allowed to work, the engineers and world of makers would find a way to help us sleep at night without interruptions from pollution or mechanical operations. If I'm wrong, then there won't be a market for this technology, and nor will there be change, because no one wants to sleep next to a power plant without a really good reason.

The Silver Lining: we can continue to improve the larger energy infrastructure, whether that means making it more efficient or replacing its resources. In this case, we should definitely consider going with maximum renewables, or a mix of renewables, NG and nuclear (I don't see any better, more mature alternatives), because we've used up all the ingenuity in the world without developing better, zero-carbon options. Again, we didn't pay for any of this change, and we shouldn't wait for results before starting to plan something else.

Worst Case Scenario #5:

The energy giants collapse. We've lost the central system.

This is a very unlikely scenario. The energy giants have been with us for a long time and they will stay with us because we need them, especially for exploration, extraction and for maintaining large national and international infrastructure (where would the NG come from? Who would operate LNG or coal terminals?).

But even without exploration or extraction, we would always need energy produced in bulk. Small- and medium-size factories, for example, have no reason to purchase their own energy units, because they would want to do what they do best – and that means making goods, not generating energy.

Many multi-unit residential buildings would not want or be able to have their own power and would prefer to stay connected to the grid. People who live in leased properties, for short periods, or people who don't care to own another piece of equipment would also need the grid.

It's much more likely that once retail market regulation is freer, the large operators will use this shift to create efficiencies and become more competitive. We will have a choice between buying or leasing our own equipment, buying our energy from a neighbour, or using the central grid, just like we can buy a car, use Uber or purchase a subway ticket. It's called competition.

But let's assume that none of this happened and we had a total collapse of the large system, with only small-to-medium-size utilities providers remaining. Most people will fear the loss of jobs. This may occur and we shouldn't dismiss it as trivial. But job losses are the inevitable result of any technological change and job security has rarely stood in the face of technology.

Where are all the people making and maintaining public payphones, telex machines, turntables, tube TVs and dial phones? Some retired, some got better jobs, and sadly, some remained unemployed and bitter. But we'll also have many more positions in the small-to-medium energy market segments, and many of the people with the technical skills can change their careers and upgrade their training to do this kind of work.

What about the facilities – those dirty and dangerous energy ghettos? The inner city ones, with their prime riverside or oceanfront locations, would be converted to museums featuring their prehistoric technologies and machinery, or they would become residential towers, commercial centres, or any other profitable use of their high-value real estate.

The rural energy plants could be converted into beautiful parklands, like the German post-industrial Landschaftspark Duisburg-Nord, which abandoned industrial activity in the 1970s and was repurposed as a post-industrial parkland in the 1990s. Some rural facilities could be converted to serve other infrastructure needs.

Large and regional transmission lines would be decommissioned. When our autonomous cars take us to the countryside (if we can ever look away from our iPhones) we'll see clear horizons and rolling hills, not 80-meter metal structures and endless strands of wires.

And what about the energy **giants** themselves, with their fancy HQs, executive lounges, mahogany tables, private jets and their users? It's very unlikely that these, along with the shareholders, would go anywhere. We would always need them and their superior know-how and equipment for resource exploration, extraction, transportation, and national infrastructure management.

The big executives and strong shareholders will know how to better position themselves in any new system architecture. They've always known. I'm sure they'll be safe in any scenario, so they shouldn't worry about this proposed future.

The Silver Lining: diversified energy system, reinvigorated economic activity, parklands, free urban space and rolling hills. It all sounds just awful.

Worst Case scenario #5:

Governments and energy giants innovate faster than the DIY market. The central system reaches zero emissions.

At this point, I can't say how probable or improbable this scenario might be, but looking back on our track record on this issue, it is does not seem so promising. Still, there's a possibility that every word in this book is wrong and the good people in government and energy corporations are actually doing a great job at innovating and bringing us closer every day to enjoying clean and cheap energy. They will eventually save our planet.

In this scenario, the wave of innovation suggested here would create many jobs and economic opportunities and a greater understanding of the importance of energy in our everyday lives, but it won't change how we consume energy on large scale. This book and its ideas have been a complete waste of everyone's time.

The Silver Lining: cynicism aside, this is actually a best-case scenario.

Worst Case Scenario #6:

Climate change doesn't occur. The science was wrong.

In this highly improbable scenario, if the ideas in this book are implemented on a large scale, we'll have a much more resilient, diverse and decentralized energy generation system and many more people will be involved in the innovation, production and trade of electric power. It's likely that this new market segment would create many economic opportunities and increased tax revenues.

The Silver Lining: **the energy giants could continue to operate freely and emit all the carbon in the world, and probably continue to provide the lion's share of our energy needs. I'm sure they'll be relieved. We've avoided spending a fortune on large-scale renewable infrastructure, at a great cost to our rural and natural environments.**

4. CONCLUDING REMARKS

This book was written over a period of few weeks between the end of 2015 and early 2016. It wasn't until I released the first Make Energy Cool post, on New Year's Eve 2015, that I realized I wanted to write this book. But as I published the post, and as I started to see the reactions from around the globe, the topic inspired me to finish writing in the shortest period possible.

I have knowingly chosen urgency and straightforwardness over rigour, because I think energy isn't a grim issue and it shouldn't be so complicated to understand, but it is *urgent*. My hope is that we can start to discuss and represent energy and climate issues in a broader cultural context, not just as pure techno-economic problems. We could have an energy culture, just as there's a car culture and a computer culture.

This book has offered a series of anecdotal views about what made us so dependent on our energy system on one hand, and so detached and uninspired by it on the other hand.

From an historical perspective, the energy industry's achievements over the last one-and-a-half centuries are very unique. With little to no marketing efforts, we all consume energy on a regular basis and our dependency on it keeps growing. With very incremental innovation, the industry has been able to continually support our needs, as our standards of living raised from those of scattered agrarian societies to the residents of a hyper-connected global village. And even though the industry has had a terrible environmental impact, contrary to common belief, many of its members didn't run away from their responsibility to alleviate some of these effects.

We should be thankful to live in an era where electric energy is everywhere, while it is still affordable. Not even Edison could have foreseen this.

But my main assertion is that we now have a much better understanding of the actual *environmental and social price of cheap energy*. We now see how too-big-to-mess-with some energy giants have become, and we can see how ineffective and indecisive our governments can be when it comes to changing their energy policies.

With these new insights we can and need to act. Acting is not only about protesting, nor about writing an angry letter to our delegates, and it is much more than simply complaining about how it's all going down the drain and nothing can be done.

If we leave the future of energy for *someone else* to resolve, to governments and big corporations, then it's the same as flipping a coin, because they may or may not prevail. And so far, they have a very narrow perspective on what our energy systems needs to provide (e.g. stability and affordability).

But with truly free markets, information technology, and human ingenuity, real action to make the future of energy our own is within our reach. If this future could be as easy and exciting as operating technologies like cars, mobile phones and computers, we could all participate in shaping it.

If we care about our lives and our planet, and if we want future generations to enjoy what we have today, then we can take a more proactive role, but one which won't require us to work harder, make dramatic changes to our lives or lose what we have worked so hard to achieve. We simply need to *Think Differently*, and we should start this rethinking today.

I hope this book can be a humble contribution to our collective thought process on *our* energy future.

5. ACKNOWLEDGMENTS

While some readers might find me knowledgeable on this issue, I actually learned many new things while writing this book. The most important insight was the unprecedented global concern over climate change, especially the concern among younger generations. It proves that while we seem to live in a loud, superficial global culture, we are actually getting wiser and we can see through the clutter, disguise and spin.

In hindsight, this book might make some environmental and climate activists look naive. But I was inspired to write it by the tens of thousands of faceless men and women, from every culture and continent, who have kept the climate issue alive for so many years and especially on the streets of Paris in November and December of 2015. The COP 21 and the political commitment leading to it would not have been possible without these brave people. This book is a small and humble contribution to this universal effort, not a disapproval of any kind.

I'd like to thank several people for their work, which acted as the undercurrent for the major ideas included in this book. I became much more aware of our influence on climate change through Tim Flannery's book, *The Weather Makers*. I was deeply inspired by Richard Florida's work on *The Rise of the Creative Class* and his ideas about harnessing and cultivating creativity for economic development. I was drawn into the monumental work of Tom Friedman's *Hot, Flat and Overcrowded*, which made energy and climate concerns much more exciting and personal than the current discourse of these issues. Though I disagree with many of his findings, I was challenged to reconsider my own convictions by Bjorn Lomborg's *The Skeptical Environmentalist*. Finally, I think that I've unintentionally borrowed, or at least absorbed the spirit of Paul Jarvis' writing in *Everything I know*. His work has allowed me to write more freely, and not let my inner critic kill this project and these ideas. Cheers for this, Rat Man.

I'd also like to thank my dear friend, Ariel Bino, who over lunch one day had unintentionally stirred me in the right direction – to search for my role in shaping the new energy future.

The first reader of this book was Michael Chartrand, who made the copy much smoother and accurate. From his Colombian north coast paradise, Mike also encouraged me to adopt a much more confident tone and made some key and well-placed insights and contributions.

The book you have just read owes much of its integrity and flow to its editor, the wonderful Cheri Hanson, who believed in its earliest and messiest form. Cheri was smart enough to ignore my arrogance, commit herself to a very tight schedule that spans across continents and time zones, and use her skills in making this book a much better read. I wish to thank her on your behalf as well.

Finally, I want to thank my better half, *Liat*, who always pushed me to go further, bolder and louder than where I initially set out to go. And to my two Inspiration Machines, my own Energy Generating Units, *Ben & Jordy*, who make me want to create a better world every day.

www.ingramcontent.com/pod-product-compliance
Lightning Source LLC
Chambersburg PA
CBHW020910180526
45163CB00007B/2696